Phenotypic Plasticity of Cuticular Hydrocarbons in Herbivorous Insects

Dissertation

zur Erlangung des akademischen Grades des

Doktors der Naturwissenschaften (Dr. rer. nat.)

eingereicht im Fachbereich Biologie, Chemie, Pharmazie

der Freien Universität Berlin

vorgelegt von

Dipl. Biol. Tobias Otte

aus Sömmerda

Berlin, im März 2015

Bibliographic information published by the Deutsche Nationalbibliothek

The Deutsche Nationalbibliothek lists this publication in the Deutsche
Nationalbibliografie; detailed bibliographic data are available
on the Internet at http://dnb.d-nb.de .

ISBN 978-3-8325-4047-0

Logos Verlag Berlin GmbH
Comeniushof, Gubener Str. 47,
10243 Berlin
Tel.: +49 (0)30 42 85 10 90
Fax: +49 (0)30 42 85 10 92
INTERNET: http://www.logos-verlag.de

für N

Diese Dissertation wurde am Institut für Biologie der Freien Universität Berlin in der Arbeitsgruppe Angewandte Zoologie/Ökologie der Tiere unter der Leitung von Frau Prof. Dr. Monika Hilker angefertigt.

1. Gutachterin: Prof. Dr. Monika Hilker
2. Gutachter: Prof. Dr. Joachim Ruther

Disputation am: 05. Juni 2015

This thesis is based on the following manuscripts:

Geiselhardt S., **Otte T.** & Hilker M. (2012). Looking for a similar partner: host plants shape mating preferences of herbivorous insects by altering their contact pheromones. *Ecology Letters*, 15, 971-977.

> SG, TO and MH designed research. SG and TO performed research. SG and TO analyzed data, and all authors wrote the manuscript. SG and TO contributed equally.

Otte T., Hilker M. & Geiselhardt S. (Manuscript). Phenotypic plasticity of mate recognition systems prevents sexual interference between two sympatric leaf beetle species.

> TO, MHand SG designed research. TO performed research. SG and TO analyzed data, and all authors wrote the manuscript.

Otte T., Hilker M. & Geiselhardt S. (2015). The effect of dietary fatty acids on the cuticular hydrocarbon phenotype of an herbivorous insect and consequences for mate recognition. *Journal of Chemical Ecology*, 41, 32-43.

> TO, MH and SG designed research. TO performed research. SG and TO analyzed data, and all authors wrote the manuscript.

Otte T., Hilker M. & Geiselhardt S. (Manuscript). Phenotypic plasticity of cuticular hydrocarbon profiles in insects.

> TO, MH and SG wrote the manuscript.

This research was supported by the Dahlem Centre of Plant Sciences by a grant to Tobias Otte.

Other publications in peer-reviewed journals:

Geiselhardt S., **Otte T.** & Hilker M. (2009). The role of cuticular hydrocarbons in male mating behavior of the mustard leaf beetle, *Phaedon cochleariae* (F.). *Journal of Chemical Ecology*, 35, 1162-1171.

Blenn B., Bandoly M., Kuffner A., **Otte T.**, Geiselhardt S., Fatouros N.E. & Hilker M. (2012). Insect egg deposition induces indirect defense and epicuticular wax changes in *Arabidopsis thaliana*. *Journal of Chemical Ecology*, 38, 882-892.

Table of Contents

1

General Introduction and Thesis Outline

The overwhelming diversity of herbivorous insects is the result of isolation barriers and evolutionary adaptation (Berlocher & Feder 2002; Drès & Mallet 2002; Matsubayashi et al. 2010; Sobel et al. 2010). Many recent studies analyzed ongoing insect speciation processes in order to gain insight into the mechanisms that generate this diversity (Schluter 2001; Via 2001; Rundle & Nosil 2005). This diversity may in part be due to divergent host plant use that can lead to the formation of host-associated races (Drès & Mallet 2002). Nevertheless, the basic process of speciation in herbivorous insects is poorly understood. The rate at which speciation proceeds, and the underlying genetic changes and ecological conditions are poorly known. Therefore, it is crucial to identify the genetically and/or phenotypically based barriers of gene flow since reproductive isolation is regarded as a key step in speciation. Furthermore, knowledge on how these barriers evolve within and between populations is essential to understand speciation processes.

The present thesis examines the chemoecological basis of behavioral isolation within and between leaf beetle populations. This behavioral isolation arises and is maintained after a shift to a novel host plant and might act as initial step of a speciation process. There are some characteristics that render leaf beetles (Chrysomelidae) an excellent model taxon to study speciation processes and the underlying mechanisms. The huge diversity of Chrysomelidae specialized on a wide range of host plants (Jaenike 1990; Funk et al. 2002) offers an ideal basis to study the mechanisms of adaptive evolution. Furthermore, even highly specialized chrysomelids are able to switch to novel host plant species leading to disruptive selection pressure between the chrysomelid populations on the ancestral and the novel host plant (Funk 1998; Berlocher & Feder 2002). The resulting ecological divergence might promote reproductive isolation barriers and reduce the gene flow between populations associated with different host plants.

1.1 Ecological speciation

The crucial step for the divergence of populations is the formation of barriers that impair gene flow between them. Theories about speciation differ in how such isolation barriers may have evolved. Most studies of speciation use geographical perspectives to classify modes of speciation. When geographical barriers (e.g. rivers or deserts) prevent exchange between populations, gene flow between these populations is interrupted, and genetic differences accumulate over time, leading to two genetically distinct populations (Coyne & Orr 2004; Butlin et al. 2008). In this classical speciation model (allopatric speciation), isolation is based on extrinsic barriers.

In contrast, in sympatric speciation scenarios, gene flow is prevented by the evolution of intrinsic barriers. In this speciation mode, divergence of populations starts without a physical/geographical barrier (Coyne & Orr 2004; Bolnick & Fitzpatrick 2007).

Regardless of allopatric or sympatric, speciation is currently considered to be driven by four potential mechanisms: (1) genetic drift, (2) divergence under uniform selection, (3) polyploidy and/or (4) ecological barriers (Schluter 2001). Speciation by genetic drift is based on genetic and/or sexual incompatibilities (Schluter 2001). Speciation by divergence under uniform selection is caused by different advantageous mutations in diverging populations inhabiting similar environments (Turelli et al. 2001), while polyploidy drives speciation due to multiplication of the entire set of chromosomes (result of hybridization) (Schluter 2001). These three mechanisms of speciation are referred to as "non-ecological" speciation modes (Sobel et al. 2010). The fourth mode of speciation is caused by divergent natural selection and has recently become of great interest: *ecological speciation*.

Ecological speciation is defined as a process by which barriers to gene flow evolve as a result of ecologically/environmentally based divergent selection on traits (Schluter 2001; Rundle & Nosil 2005; Matsubayashi et al. 2010; Sobel et al. 2010). This divergent selection is often based on differences in resources used by the diverging (parts of) populations, e.g. differences in host plants (Nosil et al. 2002). Further factors contributing to ecological speciation include sexual selection and ecological interactions (Rundle & Nosil 2005). These barriers to gene flow can arise before (prezygotic) or after (postzygotic) zygote formation (The Marie Curie SPECIATION Network 2012). Prezygotic isolation prevents gametes from encountering each other, either pre- or post-mating. Premating isolation barriers may be caused by sexual isolation (mate discrimination), adaptations to different environments (habitat isolation) or divergent developmental schedules (temporal isolation). Postzygotic isolation (genetic incompatibilities causing reduced viability or fertility of hybrids) might also occur and may take longer to evolve (Matsubayashi et al. 2010; Sobel et al. 2010).

An especially relevant form of ecological speciation is based on differences in mate choice among populations. If these differences result in assortative mating, this may initiate speciation (Rundle & Nosil 2005; Symonds & Elgar 2008; Via 2009; Cornwallis & Uller 2010; Hoskin & Higgie 2010; Maan & Seehausen 2011; Smadja & Butlin 2011). Knowledge about the mechanism that generates divergence of mating systems among populations is a key factor to understand how reproductive isolation barriers are established. Colonization of a novel host can promote differences in mate signaling and recognition, leading to divergent selection, assortative mating and potentially cause sexual isolation (Funk et al. 2002; Nosil et al. 2007; Cocroft et al. 2010). Differences in insect communication systems may influence the success of specific mate recognition signals promoting reproductive isolation (Landolt & Phillips 1997; Rundle & Nosil 2005; Smadja & Butlin 2009). Insect communication systems will diverge according to the sensory drive hypothesis, if the systems are adapted to the local habitat and if the habitats of populations differ (Boughman 2002; Maan & Seehausen 2011).

Divergence of populations and speciation processes may be associated with character displacement and reinforcement of trait development (Servedio & Noor 2003; Hoskin & Higgie 2010; Servedio et al. 2011). Character displacement describes the phenomenon that a trait that is similar in two allopatric species (or populations) diverges due to disruptive selection if the species (populations) co-occur in the same habitat. For traits associated with resource use, this phenomenon is referred to as ecological character displacement, if the traits are relevant for reproduction, this is referred to as reproductive character displacement (Pfennig & Pfennig 2009; 2012 and references therein). Reinforcement is actually a special case of character displacement which is the accentuation of differences between species by selection against the individuals of similar phenotype (reinforcement = reproductive character displacement which is achieved by selection against hybrids) (Servedio & Noor 2003; Matute 2010). The evolution of reproductive barriers via reinforcement is expected to evolve in regions where the ranges of two species overlap and hybridize as an evolutionary solution which helps avoiding the costs of maladaptive hybridization.

1.2 Mate recognition and chemical communication

Successful chemically mediated mating requires both a chemical mating signal released by the sender and the sensory system of the receiver. When both parts are inherited independently, they are under reciprocal stabilizing selection, as the signal must match the sensory properties of the receiver and *vice versa*. Insects benefit from accurate mate recognition, which occurs when the sexual signals sent by a signaler closely matches the mate preferences of the receiver (Butlin & Richtie 1994).

Most speciation models are based on the assumptions that mate signaling and recognition are under tight genetic control (Coyne 1992; Arbuthnott 2009). However, the impact of phenotypic plasticity on mate signaling and recognition is almost neglected (Cornwallis & Uller 2010). Reproductive character displacement describes the process by which mate recognition systems diverge as a response to maladaptive reproductive interactions with heterospecifics in sympatry (Pfennig & Pfennig 2009). If two species with similar mating signals co-occur, those individuals that exhibit reproductive traits most dissimilar to those of heterospecifics are favored by selection and gain a fitness benefit (Servedio & Noor 2003; Coyne & Orr 2004).

Apart from acoustic, tactile and visual stimuli, chemicals play an important role in communication systems of numerous species (Wyatt 2003 and references therein). Chemical communication based on pheromones has been studied for a wide range of species including bacteria (Ben Jacob et al. 2004), fungi (Casselton 2002), and especially invertebrate and vertebrate animals (Wyatt 2003). Insects represent the group of animals with a highly diverse range of pheromones (e.g.: Hymenoptera: Ayasse et al. 2001; Lepidoptera: Ando et al. 2004; Coleoptera: Francke & Dettner 2005). According to their function, pheromones are subdivided into, e.g., alarm pheromones (Byers 2005), trail marking pheromones (Steinmetz et al. 2003), aggregation pheromones (Torto et al. 1994) and sex pheromones (Ferveur 2005).

Production of a sex pheromone is known for either gender. Whether the female or male is the pheromone sender depends on the species considered (Johansson & Jones 2007). Behavioral changes can be elicited by single pheromone substances (Ginzel et al. 2006) or complex multi-component mixtures (Stevens 1998). However, small chemical changes reduce the effectiveness of these pheromone mixtures (Seabrook 1977). Due to their high specificity, pheromones can even function in very low concentrations (Dusenbery 1992). The distances over which pheromones are active vary from several hundred meters (Dusenbery 1992), over a few centimeters (Edwards & Seabrook 1997) to the effect only upon direct contact (Tregenza & Wedell 1997).

Divergence in the chemistry of sex pheromones of herbivorous insects can evolve as a result of isolation of populations promoting reproductive isolation and thus speciation (Smadja & Butlin 2009). While differences in sex pheromones of allopatric populations are known as pheromone dialects (Vereecken et al. 2007), less is known about variation of pheromones within a population and its pheno- or genotypic origin that can affect mating success and mate recognition (Takanashi et al. 2005). When insects acquire host plant chemicals to use them as sex pheromones or as precursors for their sex pheromone biosynthesis, a host plant dependent phenotypic change of mating signals might result in sexual isolation, and thus promote reproductive isolation and speciation (Landolt & Phillips 1997). In the lepidopteran genus *Heliothis*, strains with different host plant preferences produce different pheromone blends leading to variation in male responses (Groot et al. 2009). In *Drosophila mojavensis*, different larval rearing substrates alter the cuticular hydrocarbon profiles in adult flies (Etges et al. 2006). These cuticular hydrocarbons are used as contact cues in mate recognition, and different cuticular hydrocarbon profiles lead to sexual isolation between flies originating from different host plant lines. Similar effects were observed in *D. serrata* that were kept in different environments during larval development in the laboratory (Rundle et al. 2005).

1.3 Cuticular hydrocarbons (CHCs)

The cuticle of insects is coated with a thin lipid layer that prevents desiccation (Gibbs 1998) and protects the insect against microorganisms (Herzner & Strohm 2007), but has adopted a secondary function in chemical communication. The cuticular layer is comprised of a complex mixture of long-chain hydrocarbons with minor admixtures of other lipid classes (e.g. fatty acids, alcohols, esters, ketones, and aldehydes). Especially the CHCs can play important roles in the chemical communication system within and between insect species.

1.3.1 Function of CHCs for chemical communication

A major function of CHCs in insects is to serve as cues that mediate intra- and interspecific communication in social and solitary insects (Howard & Blomquist 2005). In some cases, individual components are recognized and induce a specific behavior, whereas in other cases a complex mixture of hydrocarbon components is necessary to elicit a specific behavior. CHCs are crucial for nestmate recognition (Ruther et al. 2002) and recognition of sexual partners in several taxa, e.g. in flies (Ferveur 2005), bees (Mant et al. 2005) and beetles (Sugeno et al. 2006). In ants, CHCs can provide information about the fertility or reproductive status (Liebig et al. 2000) in addition to their function as alarm (Loefqvist 1976) or aggregation (Bartelt et al. 1986) pheromones. Furthermore, these compounds are used for chemical mimicry (Dettner & Liepert 1994), as primer pheromones (Heifetz et al. 1997) and task-specific cues (Greene & Gordon 2003), and can even be used as anti-aphrodisiac (Oppelt & Heinze 2009).

Insect CHCs and their role as contact sex pheromones have been studied in different taxa such as Coleoptera (Zhang et al. 2003), Hymenoptera (Ruther et al. 2011), and Diptera (Etges & Jackson 2001). Different substance classes can elicit copulatory behavior in these taxa. Chemical analysis identified unsaturated hydrocarbons like (Z)-7-heptacosene (Boeroeczky et al. 2009) or (Z)-9-nonacosene (Ginzel et al. 2006) or methyl-branched hydrocarbons (Spikes et al. 2010) as contact sex pheromones. In Coleoptera, CHCs are known to be used for mate and species recognition in Staphylinidae (Peschke & Metzler 1987),

5

Coccinellidae (Hemptinne et al. 1998), Curculionidae (Mutis et al. 2009), and Cerambycidae (Ginzel et al. 2003). In addition, a few chrysomelids have been reported to use cuticular hydrocarbons as sex pheromones, i.e. the Colorado potato beetle, *Leptinotarsa decemlineata* (Jermy & Butt 1991; Otto 1997), the Japanese dock leaf beetle, *Gastrophysa atrocyanea* (Sugeno et al. 2006), the blue milkweed beetle, *Chrysochus cobaltinus* (Peterson et al. 2007), and the mustard leaf beetle *Phaedon cochleariae* (Geiselhardt et al. 2009).

1.3.2 Biosynthesis of CHCs

Insects produce a complex mixture of CHCs. In most cases, their blends of CHCs consist of different substance classes like *n*-alkanes, methyl-branched alkanes and alkenes with one or more double bonds. For some of these classes different *de novo* biosynthetic pathways are necessary (Blomquist 2010; Figure 1.1). CHCs are produced by specialized cells called oenocytes that are associated with abdominal epidermal cells or the fat body (Bagnères & Blomquist 2010). Unbranched alkenes and *n*-alkanes are formed by elongation of fatty acyl-CoAs and malonyl-CoA to produce very long-chain fatty acids which are converted to the hydrocarbon by decarboxylation (Howard & Blomquist 2005 - Figure 1.1 blue column). The positions of the double bonds in unsaturated hydrocarbons are determined by a fatty acyl-CoA desaturase (Howard & Blomquist 2005). In contrast to the formation of unbranched hydrocarbons, methyl-branched alkanes (with the exception of 2-methyl-branched alkanes) and methyl-branched alkenes are synthesized by exchange of methylmalonyl-CoA in place of malonyl-CoA at specific points during chain elongation (Nelson & Blomquist 1995; Figure 1.1 brown column). In some insects, the branched amino acids valine or isoleucine are used as precursors of methylmalonyl-CoA (Dillwith et al. 1982). An exception of the biosynthetic pathway is the formation of 2-methyl-branched alkanes (Blailock et al. 1976; Blomquist 2010). These compounds are formed by the direct elongation of either valine (even-numbered carbon backbone) or leucine (odd-numbered carbon backbone) without the use of methylmalonyl-CoA (Figure 1.1 light red column). Subsequently, three processes are needed for the final formation of CHCs: elongation, reduction, and decarboxylation (Howard & Blomquist 2005).

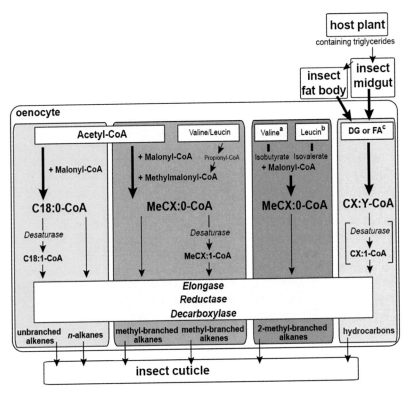

Figure 1.1: Biosynthetic pathways for the formation of different cuticular hydrocarbon classes in insects. *De novo* biosynthesis of *n*-alkanes and *n*-alkenes (blue column), internally methyl-branched saturated and unsaturated hydrocarbons (brown column) and 2-methyl-branched alkanes (light red column). Biosynthetic pathway of cuticular hydrocarbons by the elongation of dietary precursors, i.e. fatty acids (green column). For more details see description in text. [a]used for the formation of an even chain length; [b]used for the formation of an odd chain length; [c]diglycerides and free fatty acids; X = 16 to 20 carbons; Y = number of double bonds.

Besides the *de novo* biosynthesis of CHCs, insects are also known to use dietary hydrocarbons for the biosynthesis of their cuticular lipids (Blomquist & Jackson 1973; Pennanec'h et al. 1997).

In addition, triglycerides from host plants are suggested to be used by herbivorous insects as precursors for biosynthesis of their CHCs (Figure 1.1 green column). Instead of elongation of fatty acyl-CoA and malonyl-CoA, the dietary fatty acids (with a carbon backbone of C16 to C20) may be directly used as precursors for the formation of cuticular hydrocarbons. In addition to the use of dietary precursors, some hydrocarbons (like methyl-branched compounds) might still be synthesized *de novo*. After biosynthesis they are loaded onto lipophorin (a multifunctional transport protein that carries a variety of lipophilic compounds – Bagnères & Blomquist 2010), and transported to the fat body or the cuticle (Howard & Blomquist 2005).

1.4 The study system *Phaedon* (Coleoptera: Chyrsomelidae)

In this thesis, the mustard leaf beetles *Phaedon cochleariae* and *P. armoraciae* (Figure 1.2A/B) were used as model organisms to elucidate the influence of divergent host plant use on their mate recognition system. Previous studies showed that they use their CHC profiles for mate recognition (Geiselhardt et al. 2009). Both species belong to the large taxon Chrysomelidae (leaf beetles) including more than 40.000 species (Jolivet & Hawkeswood 1995; Futuyma 2004). The Chrysomelidae are closely related with the coleopteran taxa Curculionidae and Cerambycidae. Almost all leaf beetles feed on leaves, flowers, stems or roots in their larval and adult stage. The genus *Phaedon* currently includes about 75 species worldwide, 33 of which occur in Asia, seven species in Europe, 13 species in North America, 23 species in South America, one species in Africa and two species in Australia (Ge et al. 2004). These beetles mainly feed on brassicaceous plants (*P. laevigatus*, *P. brassicae*, *P. viridis*, *P. oviformis*), and many species are economically important as serious pests. However some species feed on different plant families, e.g. Asteraceae (*P. desotonis*, *P. cyanescens*), Ranunculaceae (*P. pyritosus*), Apiaceae (*P. tumidulus*) and Plantaginaceae (*P. armoraciae*).

P. cochleariae and *P. armoraciae* are widely distributed and inhabit large parts of Europe and Siberia. They produce several generations per year. Eggs are usually laid singly on the lower surface of the leaves. Larvae hatch 6 to 8 days after egg deposition, and feed on leaves, leaving the upper epidermis intact. The larval period lasts 16-17 (to 23) days with 3 molts. Third instar larvae descend to the ground and pupate in the soil. The pupal stage lasts 7 to 10 days. The adult beetle appears in the beginning of May until September. After about 14 days of maturation feeding, mating begins. A female can lay up to 400 eggs. Duration of total life span of an adult beetle is ~40 days (Koch 1992; personal observation). The pupal stage starts overwintering in October in the upper layer of the soil. The round body of adult *P. cochleariae* is dark green, nearly black, with a metallic shine. Its length varies from 2 to 3.5 mm. *P. armoraciae* is 3 to 4.5 mm long, but smaller specimens can occur. Their body is bright metallic blue, but completely black specimens are also sometimes found.

Figure 1.2: *Phaedon cochleariae* (A) and *P. armoraciae* (B) in their natural habitats on large bittercress (*Cardamine amara*, Brassicaceae) and brooklime (*Veronica beccabunga*, Plantaginaceae), respectively.

Both beetle species may co-occur in the same habitat (Figure 1.3A). The host plant range of *P. cochleariae* includes several members of the tribe Cardamineae (Brassicaceae), mainly watercress (*Nasturtium officinale*; Figure 1.3B) and large bittercress (*Cardamine amara*; Figure 1.3C); these plants are common species growing on banks along brooks and in marshes. In contrast, we found *P. armoraciae* preferably feeding on brooklime (*Veronica beccabunga*; Plantaginaceae; Figure 1.3D). However, both beetle species are reported as pests on various brassicaceous crops, e.g. horseradish (*Armoracia rusticana*), mustard (*Sinapis alba*) or Chinese cabbage (*Brassica rapa*) (Edwards & Heath 1964). Thus, these chrysomelid species show overlaps in the range of their host plant species, but also some differentiation.

When beetles co-occur in the same habitat, they show clear dietary resource partitioning even though the host plant species that are preferred by either species grow in close vicinity; hence, there is only marginal spatial separation between. Consequently, both species encounter one another with fairly high frequency, and need effective mechanisms that prevent reproductive interference.

Figure 1.3: Natural habitat (A) of *Phaedon cochleariae* and *P. armoraciae* in the north of Berlin, Germany (52°70'N; 13°31'E; Birkenwerder) with a mixed stand of large bittercress (*Cardamine amara*) and brooklime (*Veronica beccabunga*). The host plants of *P. cochleariae*, watercress *(Nasturtium officinale)* (B) and *C. amara* (C), and the host plant of *P. armoraciae*, *V. beccabunga* (D).

In both beetle species, mate recognition is based on CHCs which serve as contact pheromones. The CHC profiles of male and female *P. cochleariae* contain the same compounds including *n*-alkanes, monomethyl-branched alkanes, dimethyl-branched alkanes, unsaturated alkanes with one, two, or three double bonds, and monomethyl-branched alkenes (Geiselhardt et al. 2009). The chain-length of the hydrocarbons ranges from C19 to C45 (Figure 1.4A). The profiles are dominated by 2-methylalkanes and long-chained methyl-branched alkenes with even-chained carbon backbones. The most prominent peak is 2-methyloctacosane. The CHC profiles of male and female *P. cochleariae* are qualitatively similar; however, the patterns are sex-specific with respect to relative CHC quantities (Figure 1.4B; Geiselhardt et al. 2009).

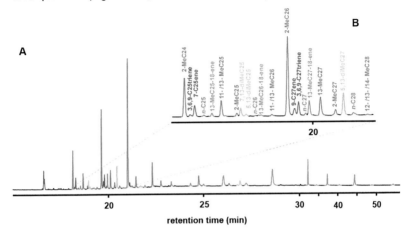

Figure 1.4: (A) Total ion current chromatogram of the cuticular hydrocarbons extracted with dichloromethane from female *Phaedon cochleariae*, feeding on large bittercress (*Cardamine amara*). (B) Detail enlargement showing the repetitive pattern of CHCs with increasing carbon chain length. Orange – *n*-alkanes; dark green – monomethyl-branched alkanes; light green – dimethyl-branched alkanes; blue – methyl-branched alkenes; red – unbranched olefins.

1.5 Outline of the thesis

The main goal of this PhD thesis is to elucidate the role of host plant use (as ecological factor) in ecological speciation of herbivorous insects. The thesis focuses on the impact of host plant use on the pattern of insect CHCs which serve as mating signals. The studies of this thesis are based on the hypothesis that feeding upon alternative host plant species leads to assortative mating by phenotypically changing the mating signals of an herbivorous insect. Such a change may lead to sexual isolation and thus, facilitate speciation processes.

All experiments were conducted with the mustard leaf beetle *Phaedon cochleariae* and the sympatric species *P. armoraciae*. As mentioned above (section 1.4), populations of both species are known to feed on brassicaceous plants, however, when both species co-occur in the same habitat they show clear resource partitioning and feed on different, co-occurring host plant species (*P. cochleariae* on brassicaceous species, *P. armoraciae* on Plantaginaceae). Hence, these two closely related chrysomelid species provide excellent model species to study the general hypothesis that ecological speciation is influenced by host plant shifts.

The thesis is divided into seven chapters, with each chapter (2 to 5) corresponding to a manuscript. Each chapter deals with a specific topic to study the above mentioned hypothesis.

In *chapter 2*, we investigated whether feeding by *P. cochleariae* on different host plant species results in assortative mating by phenotypically altering their CHC profiles, i.e. traits that are known to be involved in mate recognition. A laboratory stock of *P. cochleariae* was used to investigate this question. We found that males of the mustard leaf beetle preferred to mate with females that were reared on the same plant species to females provided with a different plant species, based on divergent CHC profiles that serve as contact pheromones. Furthermore, chemical analyses were conducted to show that the CHC phenotypes were specific for the host plant species that the beetles had fed upon. The beetles changed their CHC profiles within 14 days after a shift to a novel host plant species.

Chapter 3 elucidates if assortative mating mediated by host-induced phenotypic divergence of traits involved in mate choice still functions across species boundaries. Even though *P. cochlearieae* and *P. armoraciae* share a common host plant range, we found them in the field always feeding upon different, co-occurring plant species. When the two beetle species fed on different host plant species, they showed behavioral isolation. However, when reared on a common host species, males of both species mated randomly without distinguishing between conspecific and heterospecific females. Correspondingly, CHC phenotypes showed significant differences between both beetle species when they fed on different host species, but CHC phenotypes converged in case of similar host use. Thus, the behavioral isolation is mediated by host-induced phenotypic plasticity rather than by genetic divergence of their mate recognition systems. Our results illustrate the role of phenotypic plasticity of mate recognition systems in species discrimination and in coexistence of closely related, sympatric species.

In chapter 2 and 3 we could show that host plants can influence the chemical composition of CHC profiles of the two studied *Phaedon* species. Therefore, the next chapter (*chapter 4*) addressed the question if diet-derived compounds are directly incorporated into the insect´s CHCs, thus leading to phenotypically plastic CHC profiles. We confirmed that beetles that changed the host plant species altered their CHC pattern within one generation. Indeed, the alteration of CHC profiles was induced by incorporation of diet-derived fatty acids used as precursors for the biosynthesis of the corresponding hydrocarbons. Furthermore, we observed phenotypic changes in CHC profiles within a generation when beetles fed on different artificial diets. Our results show that a host plant shift can induce phenotypic changes in CHC profiles of an herbivorous insect by incorporating dietary compounds.

Chapter 5 is a theoretical approach which summarizes and discusses the knowledge on the phenotypic plasticity of insect cuticular hydrocarbons and the abiotic and biotic factors that influence these profiles.

The results of all four chapters (2 to 5) are summarized in *chapter 6* and *chapter 7*.

1.6 References

Ando T., Inomata S. & Yamamoto M. (2004). Lepidopteran sex pheromones. *Topics in Current Chemistry*, 239, 51-96.

Arbuthnott D. (2009). The genetic architecture of insect courtship behaviour and premating isolation. *Heredity*, 103, 15-22.

Ayasse M., Paxton R.J. & Tengo J. (2001). Mating behavior and chemical communication in the order Hymenoptera. *Annual Review of Entomology*, 46, 31-78.

Bagnères A.G. & Blomquist G.J. (2010). Site of synthesis, mechanism of transport and selective deposition of hydrocarbons, pp. 75-99, in G.J. Blomquist and A.G. Bagnères (eds.). Insect Hydrocarbons: Biology, Biochemistry and Chemical Ecology. *Cambridge University Press, Cambridge.*

Bartelt R.J., Schaner A.M. & Jackson L.L. (1986). Aggregation pheromones in five taxa of the *Drosophila virilis* species group. *Physiological Entomology*, 11, 367-76.

Ben Jacob E., Becker I., Shapira Y. & Levine H. (2004). Bacterial linguistic communication and social intelligence. *Trends in Microbiology*, 12, 366-372.

Berlocher S.H. & Feder J.L. (2002). Sympatric speciation in phytophagous insects: Moving beyond controversy? *Annual Review of Entomology*, 47, 773-815.

Blailock T.T., Blomquist G.J. & Jackson L.L. (1976). Biosynthesis of 2-methylalkanes in the crickets *Nemobius fasciatus* and *Gryllus pennsylvanicus*. *Biochemical and Biophysical Research Communications*, 68, 841-849.

Blomquist G.J. (2010). Biosynthesis of cuticular hydrocarbons, pp. 35-52, in G.J. Blomquist and A.G. Bagnères (eds.). Insect Hydrocarbons: Biology, Biochemistry and Chemical Ecology. *Cambridge University Press, Cambridge.*

Blomquist G.J. & Jackson L.L. (1973). Incorporation of labelled dietary *n*-alkanes into cuticular lipids of the grasshopper *Melanoplus sanguinipes*. *Journal of Insect Physiology*, 19, 1639-1647.

Boeroeczky K., Crook D.J., Jones T.H., Kenny J.C., Zylstra K.E., Mastro V.C. & Tumlinson J.H. (2009). Monoalkenes as contact sex pheromone components of the woodwasp *Sirex noctilio*. *Journal of Chemical Ecology*, 35, 1202-1211.

Bolnick D.I. & Fitzpatrick B.M. (2007). Sympatric speciation: models and empirical evidence. *Annual Review of Ecology, Evolution, and Systematics*, 38, 459-487.

Boughman J.W. (2002). How sensory drive can promote speciation. *Trends in Ecology and Evolution*, 17, 571-577.

Butlin R.K., Galindo J. & Grahame J.W. (2008). Sympatric, parapatric or allopatric: the most important way to classify speciation? *Philosophical Transactions of the Royal Society of London Series B-Biological Sciences*, 363, 2997-3007.

Butlin R.K. & Ritchie M.G. (1994). Mating behaviour and speciation, pp 43-79, in P.J.B. Slater and T.R. Halliday (eds.). Behaviour and Evolution. *Cambridge University Press, Cambridge.*

Byers J.A. (2005). A cost of alarm pheromone production in cotton aphids, *Aphis gossypii. Naturwissenschaften*, 92, 69-72.

Casselton L. (2002). Mate recognition in fungi. *Heredity*, 88, 142-147.

Cocroft R.B., Rodriguez R.L. & Hunt R.E. (2010). Host shifts and signal divergence: Mating signals covary with host use in a complex of specialized plant-feeding insects. *Biological Journal of the Linnean Society*, 99, 60-72.

Cornwallis C.K. & Uller T. (2010). Towards an evolutionary ecology of sexual traits. *Trends in Ecology and Evolution*, 25, 145-152.

Coyne J.A. (1992). Genetics and speciation. *Nature*, 355, 511-515.

Coyne J.A. & Orr H.A. (2004). Speciation. *Sinauer Associates, Sunderland.*

Dettner K. & Liepert C. (1994). Chemical mimicry and camouflage. *Annual Review of Entomology*, 39, 129-154.

Dillwith J.W., Nelson J.H., Pomonis J.G., Nelson D.R. & Blomquist G.J. (1982). A 13C-NMR study of methyl-branched hydrocarbon biosynthesis in the housefly. *Journal of Biological Chemistry*, 257, 11305-11314.

Drès M. & Mallet J. (2002). Host races in plant-feeding insects and their importance in sympatric speciation. *Philosophical Transactions of the Royal Society of London Series B-Biological Sciences*, 357, 471-492.

Dusenbery D.B. (1992). Sensory ecology: how organisms acquire and respond to information. *W.H. Freeman and Company, New York.*

Edwards C.A. & Heath G.W. (1964). The principles of agricultural entomology. *Chapman and Hall, London.*

Edwards M.A. & Seabrook W.D. (1997). Evidence for an air-borne sex pheromone in the Colorado potato beetle, *Leptinotarsa decemlineata. Canadian Entomologist*, 129, 667-672.

Etges W.J., Veenstra C.L. & Jackson L.L. (2006). Premating isolation is determined by larval rearing substrates in cactophilic *Drosophila mojavensis*. VII. Effects of larval dietary fatty acids on adult epicuticular hydrocarbons. *Journal of Chemical Ecology*, 32, 2629-2646.

Etges W.J. & Jackson L.L. (2001). Epicuticular hydrocarbon variation in *Drosophila mojavensis* cluster species. *Journal of Chemical Ecology*, 27, 2125-2149.

Ferveur J.F. (2005). Cuticular hydrocarbons: Their evolution and roles in *Drosophila* pheromonal communication. *Behavior Genetics*, 35, 279-295.

Francke W. & Dettner K. (2005). Chemical signalling in beetles. *Topics in Current Chemistry*, 240, 85-166.

Funk D.J. (1998). Isolating a role for natural selection in speciation: Host adaptation and sexual isolation in *Neochlamisus bebbianae* leaf beetles. *Evolution*, 52, 1744-1759.

Funk D.J., Filchak K.E. & Feder J.L. (2002). Herbivorous insects: model systems for the comparative study of speciation ecology. *Genetica*, 116, 251-267.

Futuyma D.J. (2004). Preface, in P. Jolivet, J.A Santiago-Blay and M. Schmitt (eds.), New developments in the biology of Chrysomelidae (p. xvii). *The Hague: SPB Academic Publishing*.

Ge S.-Q., Yang X.-K. & Cui J.-Z. (2004). A key to the genus *Phaedon* (Coleoptera: Chrysomelidae: Chrysomelinae) from China and the description of a new species. *Entomological News*, 114, 75-80.

Geiselhardt S., Otte T. & Hilker M. (2009). The role of cuticular hydrocarbons in male mating behavior of the mustard leaf beetle, *Phaedon cochleariae* (F.). *Journal of Chemical Ecology*, 35, 1162-1171.

Gibbs A.G. (1998). Water-proofing properties of cuticular lipids. *American Zoologist*, 38, 471-482.

Ginzel M.D., Blomquist G.J., Millar J.G. & Hanks L.M. (2003). Role of contact pheromones in mate recognition in *Xylotrechus colonus. Journal of Chemical Ecology*, 29, 533-545.

Ginzel M.D., Moreira J.A., Ray A.M., Millar J.G. & Hanks L.M. (2006) (Z)-9-Nonacosene - major component of the contact sex pheromone of the beetle *Megacyllene caryae. Journal of Chemical Ecology*, 32, 435-451.

Greene M.J. & Gordon D.M. (2003). Cuticular hydrocarbons inform task decisions. *Nature*, 423, 32.

Groot A.T., Inglis O., Bowdridge S., Santangelo R.G., Blanco C., Lopez J.D., Vargas A.T., Gould F. & Schal C. (2009). Geographic and temporal variation in moth chemical communication. *Evolution*, 63, 1987-2003.

Heifetz Y., Boekhoff I., Breer H. & Applebaum S.W. (1997). Cuticular hydrocarbons control behavioural phase transition in *Schistocerca gregaria* nymphs and elicit biochemical responses in antennae. *Insect Biochemistry and Molecular Biology*, 27, 563-68.

Hemptinne J.L., Lognay G. & Dixon A.F.G. (1998). Mate recognition in the two-spot ladybird beetle, *Adalia bipunctata*: role of chemical and behavioural cues. *Journal of Insect Physiology*, 44, 1163-1171.

Herzner G. & Strohm E. (2007). Fighting fungi with physics: food wrapping by a solitary wasp prevents water condensation. *Current Biology*, 17, 46-47.

Hoskin C.J. & Higgie M. (2010). Speciation via species interactions: the divergence of mating traits within species. *Ecology Letters*, 13, 409-420.

Howard R. & Blomquist G.J. (2005). Ecological, behavioral, and biochemical aspects of insect hydrocarbons. *Annual Review of Entomology*, 50, 371-393.

Jaenike J. (1990). Host specialization in phytophagous insects. *Annual Review of Ecology and Systematics*, 21, 243-273.

Jermy T. & Butt B.A. (1991). Method for screening female sex pheromone extracts of the Colorado potato beetle. *Entomologia Experimentalis et Applicata*, 59, 75-78.

Johansson B.J. & Jones T.M. (2007). The role of chemical communication in mate choice. *Biological Reviews*, 82, 265-289.

Jolivet P. & Hawkeswood T.J. (1995). Host-plants of Chrysomelidae of the world. *Backhuys Publishers, Leiden.*

Koch K. (1992). Die Käfer Mitteleuropas, Ökologie Band 3. *Goecke & Evers, Krefeld.*

Landolt P.J. & Phillips T.W. (1997). Host plant influences on sex pheromone behavior of phytophagous insects. *Annual Review of Entomology*, 42, 371-391.

Liebig J., Peeters C., Oldham N.J., Markstaedter C. & Hoelldobler B. (2000). Are variations in cuticular hydrocarbons of queens and workers a reliable signal of fertility in the ant *Harpegnathos saltator*? *Proceedings of the National Academy of Sciences of the United States of America*, 97, 4124-4131.

Loefqvist J. (1976). Formic acid and saturated hydrocarbons as alarm pheromones for the ant *Formica rufa*. *Journal Insect Physiology*, 22, 1331-46.

Maan M.E. & Seehausen O. (2011). Ecology, sexual selection and speciation. *Ecology Letters*, 14, 591-602.

Mant J., Brandli C., Vereecken N.J., Schulz C.M., Francke W. & Schiestl F.P. (2005). Cuticular hydrocarbons as sex pheromone of the bee *Colletes cunicularius* and the key to its mimicry by the sexually deceptive orchid, *Ophrys exaltata. Journal of Chemical Ecology*, 31, 1765-1787.

Matsubayashi K.W., Ohshima I. & Nosil P. (2010) Ecological speciation in phytophagous insects. *Entomologia Experimentalis et Applicata*, 134, 1-27.

Matute D.R. (2010). Reinforcement can overcome gene flow during speciation in *Drosophila. Current Biology*, 20, 2229-2233.

Mutis A., Parra L., Palma R., Pardo F., Perich F. & Quiroz A. (2009). Evidence of contact pheromone use in mating behavior of the raspberry weevil (Coleoptera: Curculionidae). *Environmental Entomology*, 38, 192-197.

Nelson D.R. & Blomquist G.J. (1995). Insect waxes, pp. 1-90, in R.J Hamilton (ed.). Waxes: Chemistry, Molecular Biology and Functions. *The Oily Press, Dundee, Scotland.*

Nosil P., Crespi B.J., Gries R. & Gries G. (2007). Natural selection and divergence in mate preference during speciation. *Genetica*, 129, 309-327.

Nosil P., Crespi B.J. & Sandoval C.P. (2002). Host-plant adaptation drives the parallel evolution of reproductive isolation. *Nature*, 417, 440-443.

Oppelt A. & Heinze J. (2009). Mating is associated with immediate changes of the hydrocarbon profile of *Leptothorax gredleri* ant queens. *Journal of Insect Physiology*, 55, 624-628.

Otto V.D. (1997). Some properties of the female sex pheromone of the Colorado potato beetle *Leptinotarsa decemlineata* Say. (Col. Chrysomelidae). *Anzeiger für Schaedlingskunde Pflanzenschutz Umweltschutz*, 70, 30-33.

Pennanec'h M., Bricard L., Kunesch G. & Jallon J.M. (1997). Incorporation of fatty acids into cuticular hydrocarbons of male and female *Drosophila melanogaster. Journal of Insect Physiology*, 43, 1111-1116.

Peschke K. & Metzler M. (1987). Cuticular hydrocarbons and female sex pheromones of the rove beetle, *Aleochara curtula* (Goeze) (Coleoptera: Staphylinidae). *Insect Biochemistry*, 17, 167-178.

Peterson M.A., Dobler S., Larson E.L., Juárez D., Schlarbaum T., Monsen K.J. & Francke W. (2007). Profiles of cuticular hydrocarbons mediate male mate choice and sexual isolation between hybridising *Chrysochus* (Coleoptera: Chrysomelidae). *Chemoecology*, 17, 87-96.

Pfennig D.W. & Pfennig, K.S. (2012). Development and evolution of character displacement. *Annals of the N.Y. Academy of Sciences*, 1256, 89-107.

Pfennig K.S. & Pfennig D.W. (2009). Character displacement: ecological and reproductive responses to a common evolutionary problem. *The Quarterly Review of Biology*, 84, 253-276.

Rundle H.D., Chenoweth S.F., Doughty P. & Blows M.W. (2005). Divergent selection and the evolution of signal traits and mating preferences. *PLoS Biology*, 3, 1988-1995.

Rundle H.D. & Nosil P. (2005) Ecological speciation. *Ecology Letters*, 8, 336-352.

Ruther J., Doering M. & Steiner S. (2011). Cuticular hydrocarbons as contact sex pheromone in the parasitoid *Dibrachys cavus*. *Entomologia Experimentalis et Applicata*, 140, 59-68.

Ruther J., Sieben S. & Schricker B. (2002). Nestmate recognition in social wasps: manipulation of hydrocarbon profiles induces aggression in the European hornet. *Naturwissenschaften*, 89, 111-114.

Schluter D. (2001). Ecology and the origin of species. *Trends in Ecology & Evolution*, 16, 372-380.

Seabrook W.D. (1977). Insect chemosensory responses to other insects, pp. 15-43, in H.H. Shorey and J.J. McKevlvey (eds.). Chemical control of insect behavior. *John Wiley, New York, London, Sydney, Toronto.*

Servedio M.R. & Noor M.A.F. (2003). The role of reinforcement in speciation: theory and data. *Annual Review of Ecology Evolution and Systematics*, 34, 339-364.

Servedio M.R., van Doorn G.S., Kopp M., Frame A.M. & Nosil P. (2011). Magic traits in speciation: magic but not rare? *Trends in Ecology & Evolution*, 26, 389-397.

Smadja C. & Butlin R.K. (2009). On the scent of speciation: the chemosensory system and its role in premating isolation. *Heredity*, 102, 77-97.

Smadja C. & Butlin R.K. (2011). A framework for comparing processes of speciation in the presence of gene flow. *Molecular Ecology*, 20, 5123-5140.

Sobel J.M., Chen G.F., Watt L.R. & Schemske D.W. (2010). The biology of speciation. *Evolution*, 64, 295-315.

Spikes A.E., Paschen M.A., Millar J.G., Moreira J.A., Hamel P.B., Schiff N.M. & Ginzel M.D. (2010). First contact pheromone identified for a longhorned beetle (Coleoptera: Cerambycidae) in the subfamily Prioninae. *Journal of Chemical Ecology*, 36, 943-954.

Steinmetz I., Schmolz E. & Ruther J. (2003). Cuticular lipids as trail pheromone in a social wasp. *Proceedings of the Royal Society B: Biological Sciences*, 270, 385-391.

Stevens I.D.R. (1998). Chemical aspects of pheromones, pp. 135-260, in P.E Howse, O.T Jones and I.D.R. Stevens (eds.). Insect Pheromones and Their Use in Pest Management. *Chapman & Hall, London, Weinheim, New York, Tokyo, Melbourne, Madras.*

Sugeno W., Hori M. & Matsuda K. (2006). Identification of the contact sex pheromone of *Gastrophysa atrocyanea* (Coleoptera: Chrysomelidae). *Applied Entomology and Zoology*, 41, 269-276.

Symonds M.R.E. & Elgar M.A. (2008). The evolution of pheromone diversity. *Trends in Ecology & Evolution*, 23, 220-228.

Takanashi T., Huang Y., Takahasi R., Hoshizaki S., Tatsuki S. & Ishikawa Y. (2005). Genetic analysis and population survey of sex pheromone variation in the adzuki bean borermoth, *Ostrinia scapulalis*. *Biological Journal of the Linnean Society*, 84, 143-160.

The Marie Curie SPECIATION Network (2012). What do we need to know about speciation? *Trends in Ecology & Evolution*, 27, 27-39.

Torto B., Obeng-Ofori D., Njagi P.G.N., Hassanali A. & Amiani H. (1994) Aggregation pheromone system of adult gregarious desert locust *Schistocerca gregaria* (Forskal). *Journal of Chemical Ecology*, 20, 1749-1762.

Tregenza T. & Wedell N. (1997). Definitive evidence for cuticular pheromones in a cricket. *Animal Behavior*, 54, 979-984.

Turelli M., Barton N.H. & Coyne J.A. (2001). Theory and speciation. *Trends in Ecology & Evolution*, 16, 330-343.

Vereecken N.J., Mant J. & Schiestl F. (2007). Population differentiation in female sex pheromone and male preferences in a solitary bee. *Behavioral Ecology and Sociobiology*, 61, 811-821.

Via S. (2001). Sympatric speciation in animals: the ugly duckling grows up. *Trends in Ecology & Evolution*, 16, 381-390.

Via S. (2009). Natural selection in action during speciation. *Proceedings of the National Academy of Sciences of the United States of America*, 106, 9939-9946.

Wyatt T.D. (2003). Pheromones and animal behaviour: Communication by smell and taste. *Cambridge University Press, Cambridge*.

Zhang A., Oliver J.E., Chauhan K., Zhao B., Xia L. & Xu Z. (2003). Evidence for contact sex recognition pheromone of the Asian longhorned beetle, *Anoplophora glabripennis* (Coleoptera: Cerambycidae). *Naturwissenschaften*, 90, 410-413.

2

Looking for a Similar Partner:
Host Plants Shape Mating Preferences of Herbivorous
Insects by Altering Their Contact Pheromones

2.1 Abstract

The role of phenotypical plasticity in ecological speciation and the evolution of sexual isolation is largely unknown. We investigated whether divergent host plant use in an herbivorous insect causes assortative mating by phenotypically altering traits involved in mate recognition. We found that males of the mustard leaf beetle *Phaedon cochleariae* preferred to mate with females that were reared on the same plant species to females provided with a different plant species, based on divergent cuticular hydrocarbon profiles that serve as contact pheromones. The cuticular hydrocarbon phenotypes were host plant specific and changed within 2 weeks after a shift to a novel host plant species. We suggest that plant-induced phenotypic divergence in mate recognition cues may act as an early barrier to gene flow between herbivorous insect populations on alternative host species, preceding genetic divergence and thus, promoting ecological speciation.

2.2 Introduction

Evolutionary success of a taxon may be measured by very different parameters such as age of the taxon, its reproductive power and ubiquity, or its ability to radiate and evolve high numbers of species. When considering the latter parameter, herbivorous insects are especially successful since they represent a huge part of terrestrial biodiversity. About 75% of the classified fauna determined are insects, and about a half of them feed on plants (Berlocher & Feder 2002; Grimaldi & Engle 2005). Speciation of herbivorous insects is closely associated with their specialization on host plant taxa (Bernays & Graham 1988; Futuyma & Moreno 1988; Jaenike 1990; Farrell et al. 1992; Egas et al. 2005). Both natural selection as well as stochastic events may lead to differentiation in host plant preference and (pre)adaptation, and thus cause a shift to a novel host plant taxon (Futuyma & Peterson 1985; Rundle & Nosil 2005; Matsubayashi et al. 2010).

Ecological speciation of herbivorous insects induced by a host plant shift is often linked with sympatric speciation. Prominent examples of sympatric speciation are provided by formation of insect races and species using different host plants (Berlocher & Feder 2002; Drès & Mallet 2002; Coyne & Orr 2004; Futuyma 2008). Sympatric speciation in herbivorous insects requires isolation mechanisms that limit gene flow between insect populations with diverging host plant preferences even in the absence of physical barriers between populations (Berlocher & Feder 2002; Drès & Mallet 2002; Bolnick & Fitzpatrick 2007; Smadja & Butlin 2011; Webster et al. 2012). Alternative host plant species represent a divergent selection pressure that favors local adaptations which are maintained by selection against migrants or hybrids (Egan & Funk 2009; Via 2009). In addition, differentially preferred host plant species may have significant and direct impact on prezygotic isolation if herbivorous insects mate exclusively on their hosts (habitat isolation) or if host species differ in their phenology (temporal isolation) (Coyne & Orr 2004).

The direct impact of environmental factors on mating traits and its contribution to speciation is widely discussed (Schluter 2001; Smadja & Butlin 2009; Maan & Seehausen 2011; Servedio et al. 2011; Fitzpatrick 2012). Behavioral isolation could evolve as a by-product of adaptation to alternative environments when the trait under divergent natural selection itself causes assortative mating ('magic trait'; Gavrilets 2004; Servedio et al. 2011). On the other hand, alternative environments can generate novel phenotypes that promote divergence among populations (West-Eberhard 2003; Pfennig et al. 2010; Fitzpatrick 2012). Different environments can directly affect traits involved in sexual communication (Smadja & Butlin 2009; Fitzpatrick 2012; The Marie Curie SPECIATION Network 2012). Thus, phenotypically plastic mating traits might directly lead to assortative mating between populations occurring in different habitats without adaptation or genetic divergence (Fitzpatrick 2012).

However, the hypothesis that divergent host use generates phenotypic plasticity in mating signals, leading to behavioral isolation has not been tested so far in herbivorous insects. Here, we tested this hypothesis in the oligophagous mustard leaf beetle *Phaedon cochleariae* (Chrysomelidae). In this species, males are known to use cuticular hydrocarbons (CHCs) as mate recognition cues (Geiselhardt et al. 2009). The host plant range of this leaf beetle species which is distributed all over Europe and spreading eastwards to Siberia, includes various Brassicaceae (e.g. mustard, cabbage, watercress, or bittercress), of which the two major host plants, watercress (*Nasturtium officinale*) and large bittercress (*Cardamine amara*), frequently co-occur in the same habitat along brook banks.

To test the impact of divergent host plant use on the phenotypic plasticity of mating signals and their role in assortative mating, we compared the mating behavior and CHC profiles of *P. cochleariae* beetles fed with Chinese cabbage or watercress. The tested beetles (42-days-old) were derived from a common laboratory stock colony on Chinese cabbage; they were separated from this colony in the pupal stage and either continued to be provided with Chinese cabbage or fed with watercress during the adult stage. First, we tested whether males discriminate between "same host plant females" and "different host plant females". To test the role of CHCs for mating preferences, we extracted CHCs from female beetles either fed with Chinese cabbage or watercress and studied the response of males to glass dummies treated with cuticular extracts of either "same host plant females" and "different host plant females". Additionally, we chemically compared the mating signals, i.e. CHC profiles of (male and female) adult *P. cochleariae* kept on watercress or Chinese cabbage by coupled gas chromatography–mass spectrometry. Finally, we investigated how long a beetle needs to feed on a novel plant until its CHC profile has reached a pattern that is characteristic for the novel host species. Our results showed that the plant on which a beetle feeds can shape the beetles´ mating cues, i.e. their CHC patterns, within a short time period (2 weeks); male beetles preferred mating with those females that fed on the same plants as they did. Hence, our data strongly support the hypothesis that use of different host plants triggers behavioral isolation in herbivorous insects by the host plant´s impact on the insect mating signals, thus promoting reproductive isolation and speciation of herbivorous insects.

2.3 Materials and Methods

Beetles and Plants. Adults of *Phaedon cochleariae* fed with different host plant species were studied. All adults originated from the same laboratory stock population reared on Chinese cabbage (CC) (*Brassica rapa* ssp. *pekinensis*) for at least 20 generations.

In order to avoid bottleneck effects, we took (i) a subset of 1,000 randomly chosen pupae from this stock population and continued to feed the adults emerging from these pupae with CC and (ii) a subset of further 1,000 randomly chosen pupae that were fed with watercress (WC) during the adult stage. Beetles were kept in a climate chamber at 20°C, 70% relative humidity, and a 16:8 h light:dark cycle. After eclosion of adults, beetles were separated by sex and batches of 100 beetles of the same sex were kept together in plastic containers (20 x 20 x 6.5 cm) containing moist paper towels and food (leaves cut from greenhouse-grown host plants). Beetles used for mating trials and chemical analysis were all 42 days old (since eclosion from pupae).

The plants were grown in a greenhouse with long day conditions, controlled by sodium vapor lamp SON-T AGRO 400 (photoperiod 16 h, temperature 22±4°C, relative humidity 35±10%, photosynthetic photon flux 175 µmol m^{-2} sec^{-1} across the plants). Chinese cabbage and watercress seeds were obtained from Saatzucht Quedlinburg GmbH (Quedlinburg, Germany) and Carl Sperling & Co. GmbH (Lüneburg, Germany), respectively. Beetles were fed with leaves of 6-7 week-old plants.

Mating Bioassays. Mating bioassays were conducted in Petri dishes (5.5 cm x 1.2 cm) lined with filter paper. All tests were done between 13:00 and 15:00 h CET at 22–24°C. Petri dishes were illuminated by a 60 W bulb placed 40 cm above the dishes; no daylight or other light was available in the bioassay room. Bioassays were started by placing a test male into a dish containing a potential mate or a female glass dummy (see below), and we recorded whether or not the test male started a copulation attempt (mounting the female; eversion of aedeagus) within 10 min. We conducted a no-choice bioassay, as this reflects best the natural situation where a male makes contact with only one potential mate at a given time and then decides to accept or decline the encountered object for mating.

To study male mating behavior as a function of the host plant consumed by a potential mate the following pair combinations were tested: both males and females reared on the same host plant (WC or CC) or the two reciprocal combinations (CC x WC and WC x CC). Each combination was replicated 60-times.

To investigate the role of CHCs for male mate recognition, bioassays were performed with the same four combinations, but instead of live females, glass beads (4 x 3 mm; dark green) treated with a cuticular extract of "same host" or "different host" females were offered to test males (amount per bead equivalent to 1 female; SI Materials and Methods).

In order to detect differences in male mating preferences, we compared the numbers of male mating attempts with "same host plant" and "different host plant" females and dummies, respectively, by means of the χ^2 test. Furthermore, we estimated the overall sexual isolation, I_{PSI} index, between beetles fed with CC or WC using JMATING 1.0.8 software (Carvajal-Rodríguez & Rolán-Alvarez 2006). Values of I_{PSI} range from −1 (complete disassortative mating) to 0 (random mating), to +1 (complete assortative mating), standard deviations and test of significance for I_{PSI} were obtained by bootstrapping (10,000 bootstrap samples).

Chemical Analysis of Cuticular Hydrocarbon Extracts. In order to isolate the mating signals, i.e. the CHCs, from the beetles and to analyze the dependence of the beetles' CHC pattern from the host plant, CHC extracts were prepared from male and female beetles (42-day-old) from both *P. cochleariae* fed with CC and those fed with WC during the adult stage ($N = 50$ beetles on each plant).

Prior to extraction, beetles of both sexes were killed by freezing at -18°C for 30 min and thawed at room temperature for 15 min. Extracts were prepared by extracting an individual beetle in 125 μl dichloromethane for 10 min. For quantification, 10 μl n-eicosane solution (0.1 mg/ml in hexane) was added as an internal standard. The samples were analyzed by gas chromatography coupled to mass spectrometry (SI Materials and Methods). Peak areas relative to total peak area were computed for each CHC, and peaks with a mean relative proportion of less than 0.1% were discarded from further analyses. Prior to multivariate statistics, the CHC data were centered logratio transformed as follows: $z_{ip} = ln[A_{ip}/g(A_p)]$, where A_{ip} is the area of peak i for beetle p, $g(A_p)$ is the geometric mean of all peaks for beetle p, and z_{ip} is the transformed area of peak i for beetle p (Aitchison 1986). As the logarithm is not defined for zero values, the constant 0.01 was added to each relative peak area to apply the transformation formula also to samples that did not contain all compounds (peaks).

Quantitative differences between the CHC profiles of male and female *P. cochleariae* adults fed with either CC or WC were statistically analyzed by a MANOVA with sex and host plant as main effects. Furthermore, a canonical discriminant analysis was performed to determine the effect of sex and host plant on the CHC profiles. The quality of the resulting classification was tested by "leaving-one-out" cross-validation implemented in SPSS 17. Differences between the squared Mahalanobis distances of CHC profiles of males to the group centroid of "same host" and "different host" females, respectively, were analyzed using the t-test of dependent samples.

Time Course of CHC Changes in Dependence of Host Plant Species. In order to investigate how long a beetle needs to feed on a novel host plant species until showing a host plant characteristic CHC profile, we first investigated the CHC profiles of 42-day-old *P. cochleariae* beetles that had spent their entire life cycle (from larvae to 42-day-old adults; >63 d) on either CC and WC, respectively (N = 40 individuals on each plant). These CHC profiles served as reference profiles for classification of a CHC profile as "typical CC" or "typical WC" profile by a discriminant analysis. The CHC profiles of these beetles were analyzed by coupled gas chromatography–mass spectrometry as described above, and a discriminant analysis of the quantitative profiles was performed to generate classification functions.

We further investigated the CHC profiles of 42-day-old male and female beetles which had spent their larval stages on CC, but were transferred from CC to WC either during the pupal stage, or 7, 14, 21, 28, and 35 days after adult eclosion, respectively. Thus, we obtained CHC profiles of 42-day-old beetles (N = 20 beetles each) that fed for 7, 14, 21, 28, 35, or 42 days of their adult stage on the novel host plant (WC). All individual CHC profiles were classified either into CC or WC profiles according to the classification functions generated by the discriminant analysis, and the percentage of individuals classified as WC beetles was computed for each feeding group.

2.4 Results

Male Mating Preference and Sexual Isolation. Adult male *Phaedon cochleariae* fed with Chinese cabbage (CC) or with watercress (WC) for 42 days showed a significant mating preference for females that fed on the same plant species for the same time period as the males did (Figure 2.1A). Thus, our data show significant sexual isolation (I_{PSI} = 0.15±0.07; N = 60, P < 0.05) between beetles feeding on different host plant species.

In order to elucidate whether the males' mating preference for "same host plant females" was due to the cuticular hydrocarbons (CHC) of these mates, male beetles were offered dummies treated with extracts of female CHCs. Their mating preferences matched those shown towards live females (Figure 2.1B). Males fed with CC or WC attempted to mate more often with the glass dummies treated with CHC extracts from "same host plant females" than with dummies treated with extracts from "different host plant females". The overall sexual isolation (I_{PSI}) was 0.18±0.08 (N = 60, P < 0.05). Hence, the mating preferences could be traced to the CHC patterns of the beetles feeding on different host plant species.

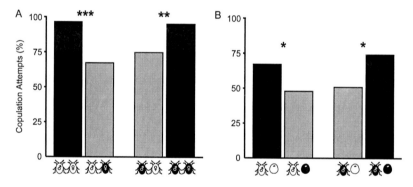

Figure 2.1: Mating responses of male *Phaedon cochleariae* towards (A) live "same host plant females" (dark grey columns) and "different host plant females" (light grey columns) and (B) glass dummies treated with cuticular hydrocarbon (CHC) extracts from these females (N = 60 for each combination). White filling = beetles fed as adults with Chinese cabbage or CHC extracts of these females. Black filling = beetles fed as adults with watercress or CHC extracts of these females. Significant differences between "same host plant" and "different host plant" combinations are indicated by *** P < 0.001; ** P < 0.01; * P < 0.05; χ^2-test.

Qualitative Chemical Composition of Cuticular Hydrocarbon Patterns. In order to detect the chemical differences in the CHC patterns of *P. cochleariae* beetles feeding on different host plant species during their adult stage, we analyzed cuticular extracts of both male and female *P. cochleariae* fed with CC or WC by coupled gas chromatography–mass spectrometry. We identified 49 peaks in these extracts (SI Table 2.1). No qualitative differences with respect to sex or host plant species of the beetles were found.

The CHC profiles comprised n-alkanes, mono- and dimethyl alkanes, unbranched olefins with one, two, or three double bonds, and monomethyl branched alkenes with chain lengths ranging from C23 to C45. All profiles were dominated by even-chained 2-methylalkanes (CC♂: 39.9±7.7%, CC♀: 36.4±8.0%, WC♂: 41.8±8.5%, WC♀: 33.5±6.2%) and long-chained methyl-branched alkenes (CC♂: 38.8±8.0%, CC♀ 37.5±8.5%, WC♂: 27.7±8.6%, WC♀: 33.9±9.7%), whereas n-alkanes represented only a small amount of the total hydrocarbons (CC♂:2.3±1.2%, CC♀: 2.3±0.8%, WC♂: 1.7±0.8%, WC♀: 1.6±0.7%).

Quantitative CHC Variation Due to Sex and Host Plant. Quantitative CHC profiles were significantly different according to sex of the beetles (MANOVA, Wilks' λ = 0.22, F = 11.17, P < 0.001) and their host plant species (Wilks' λ = 0.17, F = 15.33, P < 0.001). Additionally, we found a significant sex × host plant interaction (Wilks' λ = 0.45, F = 3.70, P < 0.001), indicating that males and females changed their quantitative CHC profile differently in response to the host plant species they were feeding upon. From the 49 peaks analyzed, 57% and 80% showed significant quantitative differences according to sex and host plant, respectively, and 29% showed a significant sex × host plant interaction (SI Table 2.1).

A discriminant analysis clearly separated males and females kept on the two host plants based on their quantitative CHC profiles (Wilks' λ = 0.02, χ^2 = 679.77, df = 144, P < 0.001) (Figure 2.2). Only 13 individuals out of 200 (6.5%) were misclassified by the original discriminant function, and 85.0% of the cross-validated cases were correctly classified. The first canonical root accounted for 58.4% of the total variance of the data and separated beetles according to their sex. The second canonical root, explaining 33.8% of the total variance, clearly separated beetles reared as adults on CC from beetles reared as adults on WC. Squared Mahalanobis distances showed that CHC profiles of males were more similar to those of "same host" females than to those of "different host" females (t-test for dependent samples, t_{WC} = -4.13, P < 0.001; t_{CC} = -12.20, P < 0.001).

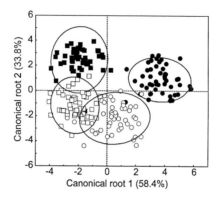

Figure 2.2: Scatterplot of canonical root 1 *vs.* 2 of a discriminant analysis based on cuticular hydrocarbons of 42-day-old *Phaedon cochleariae* males (circles) and females (squares) fed during their adult stage with Chinese cabbage (white) and watercress (black), respectively (*N* = 60 for each type).

Time Course of Changes in the CHC Pattern. In order to find out for how long a beetle needs to feed on a host plant until its CHC pattern has reached a pattern characteristic for use of this host plant, we analyzed the CHC patterns of 42-day-old *P. cochleariae* beetles that were reared for different time periods on a novel host plant species (here WC).

First, we analyzed the CHC extracts of 42-day-old beetles that fed on CC or WC during their entire life cycle (i.e. during their larval development for at least 21 d and during their adult stage for 42 d; in total at least 63 d; *N* = 40 each). The CHC profiles of these beetles were used as independent CHC reference patterns of beetles that experienced no host plant shift during their life. Based on these CHC reference patterns of beetles that fed on the same plant during their entire individual life, we performed a discriminant analysis and subsequently used the resulting classification functions to classify the CHC pattern of the test beetles into CC or WC profiles.

The CHC profiles of the reference beetles that completed their entire life cycle on the same host plant were correctly classified to 75.0% (CC; 0 d) and 92.5% (WC; >63 d) to the corresponding host plant species (Figure 2.3). When investigating the quantitative CHC profiles of beetles that experienced a host plant shift in the adult stage and fed on the novel host just for a limited period of time during the adult stage (7-42 days; *N* = 20 each), the percentages of CHC profiles classified as WC profiles varied between 40% to 100%. Interestingly, already after a 14-day-feeding period on the novel host WC, 80% of the beetles showed a WC-typical CHC profile (Figure 2.3).

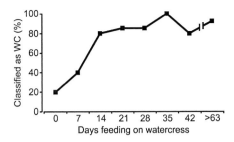

Figure 2.3: Classification of cuticular hydrocarbon (CHC) profiles of 42-day-old *Phaedon cochleariae* as "Chinese cabbage (CC)" or "watercress (WC)" profile in dependence of time the adult beetles fed on watercress (N = 20 for time points 7-42 d). CHC profiles were classified either into CC or WC profiles according to the classification functions generated by a discriminant analysis based on reference CHC profiles obtained from beetles that completed their entire life cycle on CC (0 d feeding on watercress) and WC (>63 d feeding on watercress; i.e. 42 d feeding period during the adult stage plus at least 21 d feeding period during larval stage), respectively (N = 40 for each reference type).

2.5 Discussion

Our study showed that the host plant species an herbivorous insect is feeding upon can significantly change a trait important for mate selection (here: the cuticular hydrocarbon (CHC) profile of the leaf beetle *Phaedon cochleariae*). Beetles that were fed with different host plant species (Chinese cabbage and watercress) showed quantitatively different CHC phenotypes (Figure 2.2 & 2.3). Furthermore, we found that this plant-dependent change of the beetle´s CHC phenotype leads to host plant specific assortative mating. Males preferred females that fed upon the same plant species to females that fed on the alternative plant species. This preference behavior was shown to be due to CHCs extractable from the females´ cuticle. Males preferred dummies treated with CHC extracts of "same host plant females" to "different host plant females".

We propose two explanations for the rapid divergence: (i) by external imprint of host plant CHCs on the beetles´ cuticle, or (ii) by an almost immediate impact of the host plant on the biogenesis of the CHC pattern of the beetle either by the uptake of different plant CHC precursors ingested with the different host plant species or by different plant secondary metabolites that influence the beetle´s CHC biosynthesis.

An external imprint of host plant CHCs on the beetles´ cuticle can almost be excluded as explanation for the quantitatively different CHC patterns of the beetles kept on different host. Own chemical studies of the leaf surface waxes of both CC and WC revealed that the hydrocarbon patterns of both plant species were qualitatively completely different from the beetle´s CHC patterns (unpublished data).

We favor explanation (ii) as mechanism that drives divergence of mating preferences in *P. cochleariae*. According to this explanation, the host plant alters the CHC phenotype of the beetle either by providing different precursors for the beetle´s CHC biosynthesis or by differences in secondary metabolites that interfere with the beetle´s CHC biosynthesis. The secondary metabolite profiles of CC and WC plants differ especially with respect to the presence of glucosinolates in WC and the absence of these in the CC plants used here. Whether these typical brassicaceous secondary metabolites can affect the beetle´s CHC biosynthesis is unknown. Immediate precursors for the beetle´s CHC biosynthesis could be cellular plant lipids. So far, it is unknown how CC and WC plants differ in their intracellular lipid patterns. Nevertheless, several insect species have been shown to use plant-derived fatty acids as precursors for their CHC biosynthesis (Blomquist & Jackson 1973; Pennanec'h et al. 1997), and the triacylglycerol composition in the diet is known to significantly affect CHC profiles in *Drosophila mojavensis* (Etges et al. 2006). However, the exact mechanism by which host plants affect the CHC pattern of herbivorous insects is not yet fully understood.

The divergence of mating preferences detected in *P. cochleariae* does not only require mechanisms that have caused divergence of the beetles´ CHC pattern, but also the beetles´ ability to discriminate "same host plant females" from "different host plant females". Although CHCs play an important role in insect communication (Singer 1998; Howard & Blomquist 2005), little is known about the perception and mechanisms involved in processing the information perceived from complex CHC profiles in insects. Studies on perception of CHCs for kin or nestmate recognition in social insects suggest self-referent phenotype matching ("armpit effect"; Dawkins 1982) for discrimination between close relatives and foreigners. Thereby, a self-template is compared with the profile of the counterpart (Mateo 2004). Our data suggest that *P. cochleariae* discriminates "same host plant females" from "different host plant females" by self-referent phenotype matching. The discriminant analysis has shown that CHC profiles of *P. cochleariae* depended on sex and the host plant species. The canonical root 1 separated males and females, while canonical root 2 separated the beetles according to their host plant. Based on the assumption that the male mate recognition system of *P. cochleariae* is composed of both sex-specific cues and self-referent phenotype matching, females from the same host plant match a male´s template more accurately than females from another host plant, although both types of females display the same sex-specific cues. This scenario fits the signal matching process of the sensory drive hypothesis that states that those signals are preferred that are easily detected by the receiver, i.e. that match the receiver´s receptor abilities (Endler 1992; Boughman 2002). Hence, this hypothesis may also explain how the divergence in mating traits can develop immediately after a host plant shift, even without genetic divergence.

Even though divergence of mating preferences has been demonstrated for several herbivorous insect species using different host plant species, such a rapid change of mating preferences as found in *P. cochleariae* has not explicitly been shown before. A study of the oligophagous leaf beetle *Neochlamisus bebbianae* supports our finding that mating preferences may rapidly change after host plant shifts. Populations of *N. bebbianae* using different hosts are more reproductively isolated than same-host plant populations (Funk 1998). Nevertheless, when beetles that originated from different hosts were reared on the same host plant for 1-3 weeks prior to mating trials, the sexual isolation was much weaker or even absent. On the other hand, when a subset of beetles from a same-host population was reared on an alternative host for 1-3 weeks, sexual isolation could be observed between beetles on different host plants (Funk et al. 2002). The mechanism driving this isolation has not been studied in *N. bebbianae* beetles.

However, the time necessary to trigger sexual isolation in *N. bebbianae* matches our results on the time necessary to detect CHC changes after a host plant shift by *P. cochleariae* (Figure 2.3). Our study revealed that the majority of the beetles showed a CHC pattern typical for the novel host species after only a short feeding period on the novel host during the adult stage, thus providing clear evidence for a fast phenotypic change of the CHC pattern in dependence of the host plant species used.

In the study presented here, we could link divergence of mating preferences in response to use of different host plants with a specific pattern of phenotypically plastic mating signals that may explain the divergence. Diet-induced phenotypic changes of CHCs have been reported from a few insect species (e.g. Espelie & Bernays 1989; Liang & Silverman 2000; Rundle et al. 2005; Etges et al. 2006; Sharon et al. 2010); however, an impact of diet-induced CHC profile changes on mating preference has only been shown for *Drosophila* flies (Rundle et al. 2005; Etges et al. 2006; Sharon et al. 2010). While the study on the diet-induced changes in *D. serrata* suggest that these are due to divergent selection in divergent environments (Rundle *et al.* 2005), our results demonstrated that beetles phenotypically changed their CHC pattern after having been transferred from Chinese cabbage to watercress. Sharon et al. (2010) could show that diet-inducted assortative mating in *D. melanogaster* is caused by symbiotic bacteria associated with the food media that alter the CHC phenotype. In *D. mojavensis*, the larval rearing substrate affects the CHC phenotype of the adult flies and the mating behavior. Nevertheless, in contrast to our results, *D. mojavensis* lacks host specific assortative mating, when members of the same population were reared on alternative larval substrate (Etges 1992).

A scenario on how plant-mediated phenotypic divergence of mating signals may affect genotypic divergence and thus, speciation in herbivorous insects is developed and visualized in Figure 2.4.

➤ Many models of sympatric speciation assume that divergence of different populations starts with the evolution of diverging habitat preferences followed by local adaptation, formation of different genotypes, and finally the evolution of behavioral isolation (Coyne & Orr 2004) (Figure 2.4A). In these models, habitat choice reduces migration and determines mate choice if the habitat of an herbivorous insect is characterized by the host plant and if mating takes place on the plant (Bush 1969; Diehl & Bush 1989; Via 2009). Furthermore, divergence of locally adapted populations is maintained by selection against migrants, F_1, and QTL recombinants (Via 2009). Behavioral isolation evolves as a by-product of divergent adaptations (Schluter 2001; Gavrilets 2004; Servedio et al. 2011).

➤ Unlike this classical by-product model of speciation, our data indicate that behavioral isolation may occur even without prior environmental adaptation. Our study of *P. cochleariae* revealed that feeding on different host plants can induce different CHC phenotypes which generate assortative mating. Thus, according to our data males will prefer those females which feed by chance on the same plant even when the two host plant species occur sympatrically next to each other (Figure 2.4B). The phenotypic divergence of mating signals (insect CHC pattern) is a by-product of the random host plant choice of the beetles.

➤ However, the isolating effect of host plant specific assortative mating will be neutralized by random host choice in the next generation, unless divergent host preferences evolve (Fitzpatrick 2012) (Figure 2.4C). Such host plant or habitat preferences may either be based on "preference genes" (Bush 1969; Diehl & Bush 1989; Gavrilets 2004; Via 2009), or on host plant fidelity which may be phenotypically determined (e.g. by prior experience) and become transgenerationally consistent, if females of herbivorous insects prefer those plant species for oviposition which they experienced in the juvenile phase (Maynard Smith 1966; West-Eberhard 2003; Beltman & Metz 2005). Thus, a phenotypic change of host plant preferences in an herbivorous insect can lead to assortative mating of different CHC phenotypes, reduce gene flow and may promote local adaptations and genetic divergence. Hence, in this scenario, phenotypic changes in habitat and mate preferences precede the formation of genetic divergence.

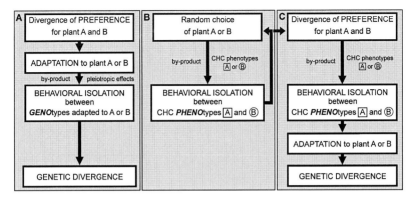

Figure 2.4: A) Classical scenario of sympatric speciation in which gene flow is mainly reduced by divergent habitat preferences and exclusive mating in the respective habitat. Behavioral isolation evolves as a by-product of local adaptation. B) Our data shows that behavioral isolation can result as a direct by-product of divergent host use due to changes in CHC phenotypes of an herbivorous insect without local adaptations. C) In combination with divergent habitat preferences, behavioral isolation may further reduce gene flow between the habitats and may precede local adaptations, thus favoring genetic divergence.

Future studies need to investigate phenotypic determination of host plant preferences in *P. cochleariae*. Furthermore, field studies need to investigate whether assortative mating also occurs in natural populations of *P. cochleariae* feeding upon different host plants that occur sympatrically. Moreover, additional studies of other plant-insect systems are needed to elucidate the prevalence of behavioral isolation based on host plant-induced divergence of insect CHC phenotypes. These studies will provide further insight into the mechanisms of sympatric speciation in herbivorous insects.

2.6 Supplemental Data

Mating Bioassays with Dummies. Extracts were prepared from freshly killed females as previously described by Geiselhardt et al. (2009). We killed females by freezing at −18°C for 30 min. After thawing for 15 min at room temperature, batches of 42 females were extracted for 10 min in dichloromethane (DCM). The extracts were concentrated under a gentle stream of nitrogen to dryness, and dissolved again in DCM. The final concentration of an extract was 0.1 female equivalent (FE) per µl.

One day (24 h) prior to bioassay start, an aliquot of 10 µl extract (= 1 FE) was applied onto a glass bead (manufacturer information: 4 x 3 mm; dark green Rocailles-drop; KnorrPrandell, Lichtenfels, Germany) which was glued onto a filter paper in the center of a Petri dish. When the bioassay was started the next day, the solvent was evaporated. We recorded whether a test male attempted to mate with a dummy (eversion of the aedeagus) within 10 min.

GC/MS Analysis. The extracts were analyzed by using a coupled gas chromatograph - mass spectrometer system (7890A GC – 5975C MSD; Agilent, Waldbronn; Germany) equipped with a cold-injection system (CIS-4, Gerstel, Mühlheim a. d. Ruhr, Germany). A volume of 1 µl extract was injected at 150°C. The CIS was immediately heated to 300°C at 12°C/s. Compounds were separated on a fused silica column (DB-5MS, 30 m x 0.32 mm ID x 0.25 µm, J & W Scientific, Folsom, USA). The helium flow was 1.2 ml/min. The oven temperature program was started at 100°C, and then heated to 300°C at a rate of 5°C/min (50 min). Electron impact ionization was 70 eV.

Hydrocarbons were identified as previously reported by Geiselhardt et al. (2009). The double-bond positions of monoenes and dienes were determined by interpreting the mass spectra of the dimethyl disulfide derivatives (Francis & Veland 1981). Methyl-branched alkenes were hydrated under a H_2 and D_2 atmosphere, respectively, using palladium on activated carbons as catalyst to determine the position of the methyl group and the relative position of the methyl group to the double bound position.

SI Table 2.1: Mean (± S.D.) relative composition (% total peak area) of cuticular hydrocarbons of male and female *Phaedon cochleariae* (N = 50) fed as adults with Chinese cabbage or watercress. Significant differences according to host (H), sex (S), and sex x host plant interaction (SH) are denoted by superscript indices for each compound (MANOVA *post hoc* Scheffé test; P < 0.05); RI = retention index

No	RI	Compound	Chinese cabbage		Watercress	
			Males	Females	Males	Females
1	2269	6,9-C23diene[S;SH]	1.1±0.7	0.9±0.7	2.9±2.0	0.7±0.6
2	2274	9-C23ene[H;S;SH]	1.8±1.1	1.4±0.9	1.5±1.0	0.7±0.4
3	2336	11-/13-MeC23[H;S]	0.1±0.1	0.0±0.1	0.2±0.1	0.1±0.1
4	2462	2-MeC24[S]	1.5±0.9	3.4±1.1	1.6±0.8	4.0±1.9
5	2479	3,6,9-C25triene / 7-C25ene[H;S]	0.3±0.2	0.7±0.4	0.2±0.1	0.4±0.2
6	2500	n-C25[H;S]	0.1±0.1	0.2±0.1	0.1±0.1	0.2±0.1
7	2507	13-MeC25-7-ene / 15-MeC25-7-ene[H;SH]	0.2±0.2	0.3±0.4	0.2±0.1	0.2±0.2
8	2534	11-/13-MeC25[S]	1.0±0.8	2.2±1.1	1.1±0.7	3.0±1.9
9	2563	2-MeC25[S;SH]	0.2±0.1	0.3±0.2	0.2±0.1	0.4±0.2
10	2573	7,13-diMeC25[H;S]	0.5±0.2	0.5±0.2	0.9±0.5	0.9±0.4
11	2582	5,13-diMeC25	0.3±0.4	0.2±0.1	0.3±0.2	0.3±0.1
12	2664	2-MeC26[H;S]	4.3±2.0	6.6±1.8	7.6±2.8	11.0±3.0
13	2676	6,9-C27diene / 9-C27ene	0.7±0.4	0.7±0.4	0.9±0.5	0.8±0.5
14	2682	3,6,9-C27triene / 7-C27ene[H;S]	0.5±0.3	0.9±0.5	0.5±0.3	0.7±0.5
15	2700	n-C27[H]	0.6±0.6	0.6±0.2	0.4±0.2	0.5±0.2
16	2707	13-MeC27-2-ene / 15-MeC27-4/9-ene[*;H]	1.8±1.3	1.7±1.0	1.4±0.8	1.7±1.9
17	2732	13-MeC27[H;S]	1.6±1.5	3.6±2.5	2.3±0.9	5.0±3.7
18	2764	2-MeC27 / 9,13-/9,15-diMeC27[H;SH]	0.8±0.8	0.9±0.7	1.1±0.3	0.9±0.5
19	2782	5,13-/5,15-diMeC27[H;S]	0.9±0.4	1.1±0.5	2.3±1.0	2.6±0.8
20	2800	n-C28[S]	0.3±0.3	0.4±0.5	0.5±0.4	0.4±0.5
21	2862	2-MeC28[S]	27.9±6.2	23.5±6.3	29.5±6.9	19.4±4.6
22	2876	6,9-C29diene[S]	1.0±0.6	0.7±0.7	1.6±1.1	0.8±0.6
23	2878	9-C29ene[H;S]	0.5±0.5	0.2±0.4	0.9±0.8	0.3±0.5
24	2900	n-C29[H]	1.2±0.6	1.1±0.5	0.6±0.5	0.5±0.5
25	2931	13-/15-MeC29[H;SH]	1.2±0.7	1.3±0.4	2.6±0.8	2.4±0.9
26	2982	5,13-/5,15-diMeC29[H]	0.5±0.3	0.4±0.4	0.3±0.2	0.2±0.3
27	3031	12-/13-/14-MeC30[H]	0.0±0.1	0.0±0.1	0.2±0.2	0.2±0.6
28	3063	2-MeC30[H;S;SH]	6.6±3.0	5.2±2.0	3.4±1.6	1.8±0.9

SI Table 2.1: continued

No	RI	Compound	Chinese cabbage		Watercress	
			Males	Females	Males	Females
29	3132	13-MeC31[H;S]	0.7±0.4	0.6±0.3	1.5±0.6	1.1±0.7
30	3205	3,13-/3,15-diMeC31[H;SH]	0.3±0.2	0.3±0.2	0.7±0.2	0.5±0.2
31	3331	13-MeC33[H;S]	0.6±0.3	0.5±0.2	1.0±0.4	0.8±0.4
32	3382	5,13-diMeC33[H]	1.0±0.4	0.9±0.3	1.7±0.6	1.5±0.5
33	3405	3,13-diMeC33[H;S]	0.3±0.2	0.2±0.1	0.5±0.2	0.4±0.3
34	3506	21-MeC35-10/12-ene / 23-MeC35-12/14-ene[H;S]	2.4±1.3	1.1±0.5	4.0±1.7	1.8±0.9
35	3532	13-/15-MeC35[H]	1.0±0.4	0.9±0.3	1.2±0.4	1.2±0.5
36	3572	7,13-/7,15-diMeC35[H]	0.1±0.1	0.1±0.1	0.1±0.2	0.2±0.3
37	3582	5,13-/5,15-diMeC35	1.1±0.4	1.0±0.3	1.3±0.7	1.3±0.5
38	3607	23-MeC36-12-ene / 24-MeC36-13-ene / 25-MeC36-14-ene[H;S]	0.8±0.2	0.6±0.2	1.1±0.4	0.8±0.3
39	3708	23-MeC37-12/14-ene / 25-MeC37-14/16-ene[H;S]	4.3±1.5	3.1±1.0	3.9±1.5	3.3±1.5
40	3731	13-/15-MeC37[H;S;SH]	0.5±0.3	0.6±0.3	0.3±0.3	0.6±0.6
41	3782	5,13-diMeC37[H]	0.2±0.2	0.2±0.3	0.3±0.3	0.3±0.3
42	3808	25-MeC38-14-ene / 26-MeC38-15-ene / 27-MeC38-16-ene[H]	1.2±1.8	1.0±0.4	0.7±0.3	1.1±0.8
43	3909	25-MeC39-14/16-ene / 27-MeC39-16/18-ene[H;S;SH]	9.2±3.9	10.2±3.9	4.2±2.3	8.4±4.2
44	4008	27-MeC40-16-ene / 28-MeC40-17-ene / 29-MeC40-18-ene[H;S;SH]	0.7±0.3	0.7±0.4	0.5±0.4	0.8±0.8
45	-†	27-MeC41-16/18-ene / 29-MeC41-18/20-ene[H;S;SH]	5.4±1.5	6.4±1.7	2.8±1.3	5.2±2.1
46	-	29-MeC42-18-ene / 30-MeC42-19-ene / 31-MeC42-20-ene[H;SH]	1.3±1.5	1.1±0.5	0.7±0.4	1.1±1.1
47	-	29-MeC43-18/20-ene / 31-MeC43-20/22-ene[H]	8.5±2.8	8.2±2.5	5.8±2.7	6.7±3.0
48	-	31-MeC44-20-ene / 32-MeC44-21-ene / 33-MeC44-22-ene[H]	0.8±0.6	0.9±0.9	0.6±0.7	0.9±1.2
49	-	31-MeC45-20/22-ene / 33-MeC45-22/24-ene[H]	2.2±0.7	2.2±0.7	1.9±0.9	1.8±0.8

* 15-MeC27-4/9-ene refers to 15-MeC27-4-ene + 15-MeC27-9-ene.
† No reference alkanes were available for the calculation of retention indices.

2.7 References

Aitchison J. (1986). The Statistical Analysis of Compositional Data. *Chapman and Hall, London.*

Beltman J.B. & Metz J.A.J. (2005). Speciation: more likely through a genetic or through a learned habitat preference? *Proceedings of the Royal Society of London, Series B: Biological Sciences*, 272, 1455-1463.

Berlocher S.H. & Feder J.L. (2002). Sympatric speciation in phytophagous insects: Moving beyond controversy? *Annual Review of Entomology*, 47, 773-815.

Bernays E. & Graham M. (1988). On the evolution of host specificity in phytophagous arthropods. *Ecology*, 69, 886-892.

Blomquist G.J. & Jackson L.L. (1973). Incorporation of labelled dietary *n*-alkanes into cuticular lipids of the grasshopper *Melanoplus sanguinipes*. *Journal of Insect Physiology*, 19, 1639-1647.

Bolnick D.I. & Fitzpatrick B.M. (2007). Sympatric speciation: models and empirical evidence. *Annual Review of Ecology, Evolution, and Systematics*, 38, 459-487.

Boughman J.W. (2002). How sensory drive can promote speciation. *Trends in Ecology & Evolution*, 17, 571-577.

Bush G.L. (1969). Sympatric host race formation and speciation in frugivorous flies of the genus *Rhagoletis* (Diptera, Tephritidae). *Evolution*, 2, 237-251.

Carvajal-Rodríguez A. & Rolán-Alvarez E. (2006). JMATING: a software for the analysis of sexual selection and sexual isolation effects from mating frequency data. *BMC Evolutionary Biology*, 6, 40.

Coyne J.A. & Orr H.A. (2004). Speciation. *Sinauer, Sunderland.*

Dawkins R. (1982). The Extended Phenotype. *W.H. Freeman, San Francisco.*

Diehl S.R. & Bush G.L. (1989). The role of habitat preference in adaptation and speciation, pp. 345-365, in D. Otte and J.A. Endler (eds.). Speciation and its Consequences. *Sinauer, Sunderland.*

Drès M. & Mallet J. (2002). Host races in plant-feeding insects and their importance in sympatric speciation. *Philosophical Transactions of the Royal Society B: Biological Sciences*, 357, 471-492.

Egan S.P. & Funk D.J. (2009). Ecologically dependent postmating isolation between sympatric host forms of *Neochlamisus bebbianae* leaf beetles. *Proceedings of the National Academy of Sciences of the United States of America*, 106, 19426-19431.

Egas M., Sabelis M.W. & Dieckmann U. (2005). Evolution of specialization and ecological character displacement of herbivores along a gradient of plant quality. *Evolution*, 59, 507-520.

Endler J.A. (1992). Signal, signal conditions, and the direction of evolution. *The American Naturalist*, 139, 125-153.

Espelie K.E. & Bernays E.A. (1989). Diet-related differences in the cuticular lipids of *Manduca sexta* larvae. *Journal of Chemical Ecology*, 15, 2003-2017.

Etges W.J. (1992). Premating isolation is determined by larval rearing substrates in cactophilic *Drosophila mojavensis*. *Evolution*, 46, 1945-1950.

Etges W.J., Veenstra C.L. & Jackson L.L. (2006). Premating isolation is determined by larval rearing substrates in cactophilic *Drosophila mojavensis*. VII. Effects of larval dietary fatty acids on adult epicuticular hydrocarbons. *Journal of Chemical Ecology*, 32, 2629-2646.

Farrell B.D., Mitter C. & Futuyma D.J. (1992). Diversification at the insect-plant interface. *BioScience*, 42, 34-42.

Fitzpatrick B.M. (2012). Underappreciated consequences of phenotypic plasticity for ecological speciation. *International Journal of Ecology*, 2012, Article ID 256017.

Francis G.W. & Veland K. (1981). Alkylthiolation for the determination of double-bond positions in linear alkenes. *Journal of Chromatography*, 219, 379-384.

Funk D.J. (1998). Isolating a role for natural selection in speciation: Host adaptation and sexual isolation in *Neochlamisus bebbianae* leaf beetles. *Evolution*, 52, 1744-1759.

Funk D.J., Filchak K.E. & Feder J.L. (2002). Herbivorous insects: model systems for the comparative study of speciation ecology. *Genetica*, 116, 251-267.

Futuyma D.J. (2008). Sympatric speciation: norm or exception? pp. 136-148, in K.J. Tilmon (ed.). Evolutionary Biology of Herbivorous Insects. *University of California Press, Berkley.*

Futuyma D.J. & Moreno G. (1988). The evolution of ecological specialization. *Annual Review of Ecology, Evolution, and Systematics*, 19, 207-223.

Futuyma D.J. & Peterson S.C. (1985). Genetic variation in the use of resources by insects. *Annual Review of Entomology*, 30, 217-238.

Gavrilets S. (2004). Fitness Landscapes and the Origin of Species. *Princeton University Press, Princeton.*

Geiselhardt S., Otte T. & Hilker M. (2009). The role of cuticular hydrocarbons in male mating behavior of the mustard leaf beetle, *Phaedon cochleariae* (F.). *Journal of Chemical Ecology*, 35, 1162-1171.

Grimaldi D. & Engle M.S. (2005). Evolution of the Insects. *Cambridge University Press, Cambridge.*

Howard R.W. & Blomquist G.J. (2005). Ecological, behavioral, and biochemical aspects of insect hydrocarbons. *Annual Review of Entomology*, 50, 371-393.

Jaenike J. (1990). Host specialization in phytophagous insects. *Annual Review of Ecology, Evolution, and Systematics,* 21, 243-273.

Liang D. & Silverman J. (2000). "You are what you eat": diet modifies cuticular hydrocarbons and nestmate recognition in the Argentine ant, *Linepithema humile. Naturwissenschaften*, 87, 412-416.

Maan M.E. & Seehausen O. (2011). Ecology, sexual selection and speciation. *Ecology Letters*, 14, 591-602.

Mateo J.M. (2004). Recognition systems and biological organization: The perception component of social recognition. *Annales Zoologici Fennici*, 41, 729-745.

Matsubayashi K.W., Ohshima I. & Nosil P. (2010). Ecological speciation in phytophagous insects. *Entomologia Experimentalis et Applicata*, 134, 1-27.

Pennanec'h M., Bricard L., Kunesch G. & Jallon J.M. (1997). Incorporation of fatty acids into cuticular hydrocarbons of male and female *Drosophila melanogaster*. *Journal of Insect Physiology*, 43, 1111-1116.

Pfennig D.W., Wund M.A., Snell-Rood E.C., Cruickshank T., Schlichting C.D. & Moczek A.P. (2010). Phenotypic plasticity's impacts on diversification and speciation. *Trends in Ecology & Evolution*, 25, 459-467.

Rundle H.D. & Nosil P. (2005). Ecological speciation. *Ecology Letters*, 8, 336-352.

Rundle H.D., Chenoweth S.F., Doughty P. & Blows M.W. (2005). Divergent selection and the evolution of signal traits and mating preferences. *PLoS Biology*, 3, 1988-1995.

Schluter D. (2001). Ecology and the origin of species. *Trends in Ecology & Evolution*, 16, 372-380.

Servedio M.R., Van Doorn G.S., Kopp M., Frame A.M. & Nosil, P. (2011) Magic traits in speciation: 'magic' but not rare? *Trends in Ecology & Evolution*, 26, 389-397.

Sharon G., Segal D., Ringo J.M., Hefez A., Zilber-Rosenberg I. & Rosenberg E. (2010). Commensal bacteria play a role in mating preference of *Drosophila melanogaster*. *Proceedings of the National Academy of Sciences of the United States of America*, 107, 20051-20056.

Singer T. (1998). Roles of hydrocarbons in the recognition systems of insects. *American Zoologist*, 38, 394-405.

Smadja C. & Butlin R.K. (2009). On the scent of speciation: the chemosensory system and its role in premating isolation. *Heredity*, 102, 77-97.

Smadja C. & Butlin R.K. (2011). A framework for comparing processes of speciation in the presence of gene flow. *Molecular Ecology*, 20, 5123-5140.

Maynard Smith J. (1966). Sympatric speciation. *The American Naturalist*, 100, 637-650.

The Marie Curie SPECIATION Network (2012). What do we need to know about speciation? *Trends in Ecology & Evolution*, 27, 27-39.

Via S. (2009). Natural selection in action during speciation. *Proceedings of the National Academy of Sciences of the United States of America*, 106, 9939-9946.

Webster S.E., Galindo J., Grahame J.W. & Butlin R.K. (2012). Habitat choice and speciation. *International Journal of Ecology*, 2012, Article ID 154686.

West-Eberhard M.J. (2003). Developmental Plasticity and Evolution. *Oxford University Press, Oxford.*

3

Phenotypic Plasticity of Mate Recognition Systems Prevents Sexual Interference between Two Sympatric Leaf Beetle Species

3.1 Abstract

Maladaptive sexual interactions among heterospecific individuals (sexual interference) can prevent the coexistence of species. Thus, the avoidance of sexual interference by divergence of mate recognition systems is crucial for a stable coexistence in sympatry. Mate recognition systems are thought to be under tight genetic control. However, we demonstrate that mate recognition systems of two closely related beetle species show a high level of phenotypic plasticity. Mate choice in these beetles is mediated by cuticular hydrocarbons (CHCs). Divergent host plant use causes a divergence of CHC phenotypes, whereas similar host use leads to their convergence. Consequently, both species exhibit significant behavioural isolation when reared on alternative host species but mate randomly when using a common host. Thus, sexual interference between these syntopic leaf beetles is prevented by host-induced phenotypic plasticity, rather than by genotypic divergence of mate recognition systems.

3.2 Introduction

The coexistence of different species in sympatry depends on mechanisms that reduce interspecific interactions with a negative fitness outcome for at least one of the species. Otherwise, one species will outcompete and displace the other ("competitive exclusion principle", Hardin 1960). Thus, species competing for limited resources will experience strong natural selection that leads to spatial, temporal, or dietary resource partitioning between them (Schoener 1974). However, the importance of resource competition in phytophagous insect communities has been the subject of controversial discussion (Strong et al. 1984;

41

Denno et al. 1995; Reitz & Trumble 2002; Kaplan & Denno 2007). Sexual interference might be another critical factor that determines the coexistence of species, especially for closely related species (Ribeiro & Spielman 1986; Kuno 1992; Gröning & Hochkirch 2008).

Reproductive interference includes any form of interspecific sexual interaction that results in fitness loss for one or both species, ranging from signal jamming during mate attraction to heterospecific matings (Gröning & Hochkirch 2008). Heterospecific mating or "satyrization" results from an overlap of the mate recognition systems of both species and the inability to discriminate conspecific and heterospecific mating cues (Ribeiro & Spielman 1986). The adverse fitness consequences of heterospecific matings are particularly high if both species are incompatible due to complete postmating isolation barriers (Liou & Price 1994). The consequences are similar to resource competition, leading either to resource partitioning or to the displacement of one species ("sexual exclusion") (Hochkirch et al. 2007; Gröning & Hochkirch 2008).

Host plant shifts achieve resource partition and may represent a key driver for ecological speciation in phytophagous insects (Futuyma & Peterson 1985; Berlocher & Feder 2002; Rundle & Nosil 2005; Matsubayashi et al. 2010; Nosil 2012). Alternative host choices may result in spatial isolation or in temporal isolation if host species differ in their phenology (Coyne & Orr 2004; Webster et al. 2012). However, in spite of alternative host choices, spatial isolation may be incomplete if a foraging insect species moves across a plant community and thereby encounters not only potential mates but also heterospecific individuals on any plant species, irrespective of their specific host species. In this situation, insects need to be able to distinguish between conspecific and heterospecific individuals. Hence, the mate recognition systems of closely related species must diverge to prevent sexual interference (Gröning & Hochkirch 2008).

Mate recognition systems consist of two components, the signal of the sender and the sensory system of the receiver. Mate recognition systems are generally assumed to be under tight genetic control (Coyne 1992; Arbuthnott 2009; Laturney & Moehring 2012). Both signals and sensory systems are thought to be inherited independently but are under reciprocal stabilizing selection because the signal must match the sensory properties of the receiver and *vice versa* (Butlin & Ritchie 1994). Hence, this reciprocal coordination of the sender and receiver may counteract the divergence of mate recognition systems. However, this genetic perspective neglects the impact of phenotypic plasticity on mate recognition systems (Green 2002; Cornwallis & Uller 2010).

Several studies have shown that environmental factors (e.g., host or diet) may significantly affect both mating signals and preferences within phytophagous insect species (Etges 1992; Funk et al. 2002; Sharon et al. 2010; Geiselhardt et al. 2012). In the mustard leaf beetle *Phaedon cochleariae*, divergent host plant use causes assortative mating by altering phenotypic traits involved in mate recognition, i.e., cuticular hydrocarbon (CHC) phenotypes (Geiselhardt et al. 2009; 2012). Males prefer to mate with females that have been reared on the same plant species to females reared on an alternative plant species; the "same host" females show CHC phenotypes more similar to those of the males than the females reared on an alternative host plant species (Geiselhardt et al. 2012). However, it is yet unknown whether host use in phytophagous insects can interfere with sexual isolation between different species.

Here, we addressed the question of whether the same mechanism that leads to host-specific assortative mating within a species can prevent reproductive interference between two closely related species with divergent host plant use, thus allowing for coexistences in (micro)sympatry. We used the two mustard leaf beetles *P. armoraciae* and *P. cochleariae*, which co-occur in the same microhabitats over large parts of their distribution area but differ in host use. In natural habitats, *P. armoraciae* feeds on brooklime (*Veronica beccabunga*; Plantaginaceae), whereas *P. cochleariae* feeds on several members of the tribe Cardamineae (Brassicaceae) (Böhme 2001). Both beetle species refuse to accept the host of the other beetle species even in no-choice experiments (unpublished data). However, in agrosystems, both species have been reported as pests on several brassicaceous crops (Edwards & Heath 1964). Despite a complete dietary resource partitioning their natural habitat, they show only weak spatial segregation. The host plant species of *P. armoraciae* and *P. cochleariae* frequently form mixed stands along brook banks and in marsh areas, and both beetle species frequently encounter one another because they regularly pass by each other's host plant when foraging for their specific hosts or for mates. Thus, they need effective mechanisms that prevent reproductive interference.

To test the impact of dietary resource partitioning on the phenotypic plasticity of mating signals and its role in interspecific behavioral isolation, we compared the mating behaviors and CHC phenotypes of both species reared either on their natural host plant species or on a common host species, i.e., Chinese cabbage (*Brassica rapa* ssp. *pekinensis*).

3.3 Materials and Methods

Beetles. Adults of *Phaedon cochleariae* and *P. armoraciae* were field-collected on *Cardamine amara* (C) (Brassicaceae) and *Veronica beccabunga* (V) (Plantaginaceae), respectively, in Birkenwerder, a locality 30 km north of Berlin, Germany. The species were kept separately in a climate chamber at 20°C, 70 % relative humidity, and a 16:8 h light:dark cycle. Beetles of both sexes were placed together in a container (20 x 20 x 6.5 cm) containing moist paper towels. They were provided with leaves of greenhouse-grown plants. Each beetle species was fed with the plant species upon which it had been collected in the field (i.e., its native plant species). To investigate the effect of host plant use on interspecific behavioral isolation, we generated two different lines of each beetle species that were either fed with their native host species or with Chinese cabbage, *Brassica rapa*. ssp. *pekinensis* (B), (Brassicaceae). Subsequently, egg-laden leaves were transferred into a new container where larvae hatched and were fed with leaves of the respective host plant species. All individuals used for mating trials and chemical analysis were the F_1 progeny of field-caught beetles.

Mating Bioassays. Beetles used for mating trials and chemical analyses were separated by sex within the first 7 days after eclosion and were maintained separately until the beetles were 21 days old to ensure sexual maturity. The mating bioassays were conducted in Petri dishes (5.5 cm x 1.2 cm) lined with filter paper. All tests were performed between 13:00 and 15:00 h CET (7-9 h after light was switched on) at 22–24 °C. Petri dishes were illuminated with a 60 W bulb placed 40 cm above the dishes; no daylight or other light was available in the bioassay room. Bioassays were started by placing a test male into a Petri dish containing a potential mate or a dummy. We recorded whether the test male started a copulation attempt within 10 min and, if so, how long the copulation lasted.

To study the impact of host plant use on interspecific behavioral isolation, the following male×female mating combinations were tested: (i) males and females that were fed on their natural host species (PcC×PcC, PaV×PaV, PcC×PaV, PaV×PcC); and (ii) males and females that had fed on Chinese cabbage (PcB×PcB, PaB×PaB, PcB×PaB and PaB×PcB). Forty pairs were used for each combination.

The differences in the number and duration of intra- and interspecific mating attempts were analyzed with χ^2-tests and Mann-Whitney U tests, respectively. To investigate the role of cuticular hydrocarbons (CHCs) in male mate recognition, the bioassays were performed with the same mating combinations as described above, but instead of females, glass beads (manufacturer information: 4 x 3 mm; dark green Rocailles-drop; KnorrPrandell, Lichtenfels, Germany) treated with 1 female equivalent of the respective cuticular extract were offered to test males (Geiselhardt et al. 2009).

Extraction of CHCs. For the preparation of CHC extracts for use in bioassays, females were killed by freezing them at -18 °C for 30 min. After thawing for 15 min at room temperature, batches of 42 females were extracted in dichloromethane (DCM) for 10 min. The extracts were concentrated under a gentle stream of nitrogen to dryness and were dissolved in DCM. The final concentrations of the extracts were 0.1 FE per μl.

For the chemical comparison of CHC phenotypes, CHC extracts of males and females of both host plant lines of each species ($N = 40$ each) were prepared by extracting individual beetles in 125 μl of DCM for 10 min. For quantification, 10 μl of an *n*-eicosane solution (0.1 mg/ml in hexane) was added as an internal standard.

Chemical analysis. The samples were analysed using a coupled gas chromatograph-mass spectrometer system (7890A GC – 5975C MSD; Agilent, Waldbronn; Germany). A volume of 1 µl of extract was injected at 300 °C. The compounds were separated on a fused silica column (DB-5MS, 30 m x 0.25 mm ID x 0.25 µm, J & W Scientific, Folsom, USA). The helium flow was 1 ml/min. The oven was programmed to rise in temperature from 100 °C to 320 °C (35 min isotherm) at 10 °C/min. The electron impact ionization was 70 eV.

Hydrocarbons were identified as previously reported by Geiselhardt et al. (2009). For quantification of CHCs, the peak areas of each CHC relative to the total peak area were calculated. Peaks close to the detection limit (i.e., with a mean proportion of less than 0.1 %) were discarded from further analysis. Prior to multivariate statistics, the CHC data were centred log-ratio transformed as follows: $z_{ip} = \ln[A_{ip}/g(A_p)]$, where A_{ip} is the area of peak *i* for beetle *p*, $g(A_p)$ is the geometric mean of all peaks for beetle *p*, and z_{ip} is the transformed area of peak *i* for beetle *p* (Aitchinson 1980). As the logarithm is not defined for zero values, the constant 0.01 was added to each relative peak area to apply the transformation formula also to samples that did not contain all compounds (peaks).

Statistical analysis. The magnitude of behavioural isolation between both species reared either on their native host or on Chinese cabbage, was estimated by the overall sexual isolation, or I_{PSI} index, using JMATING software (Rolán-Alvarez & Caballero 2000; Carvajal-Rodríguez & Rolán-Alvarez 2006). The values of I_{PSI} range from −1 to +1, with $I_{PSI} > 0$ indicating sexual isolation between both species. The standard deviations and tests of significance for I_{PSI} indices were obtained by bootstrapping (10,000 bootstrap samples) implemented in the JMATING software.

Quantitative differences between the CHC profiles of *P. cochleariae* and *P. armoraciae* lines were analysed by a canonical discriminant analysis to determine the effects of species and host on the CHC profiles. The quality of the resulting classification was tested by "leaving-one-out" cross-validation implemented in SPSS 17.

3.4 Results

Sexual isolation when using different host plant species. The two mustard leaf beetles *Phaedon armoraciae* (Pa) and *P. cochleariae* (Pc) were reared in the laboratory on *Veronica beccabunga* (V) and *Cardamine amara* (C), respectively, which correspond to the host plant species on which these beetles were found feeding upon in the field. Under these natural conditions, both *Phaedon* species showed a significant degree of interspecific sexual isolation (PaV-PcC: I_{PSI} =

0.61 ± 0.09; $P < 0.001$). The males of both species mated significantly longer (PaV: $U = 3$; $P < 0.002$; PcC: $U = 138$, $P < 0.023$; Figure 3.1A) and more often (PaV: χ^2 = 39.60, $P < 0.001$; PcC: $\chi^2 = 20.83$; $P < 0.001$; Figure 3.1B) with conspecific females compared to heterospecific females. In more than 90 percent of the cases, intraspecific mating lasted more than 20 min, whereas all *P. armoraciae* males and half of the *P. cochleariae* males aborted interspecific copulation attempts before this time.

When dummies treated with CHC extracts of females were offered to males, the male mating preferences matched those shown towards live females (Figure 3.1C). The males of both species attempted to mate approximately 3 times more often with glass dummies treated with CHC extracts from conspecific females than with dummies treated with extracts from heterospecific females (PaV: χ^2 = 14.7, $P < 0.001$; PcC: $\chi^2 = 15.2$, $P < 0.001$). The overall sexual isolation (I_{PSI}) explained by CHCs was 0.55 ± 0.11 ($P < 0.001$).

No sexual isolation when using the same host plant species. When both species were reared on the same host plant (*Brassica rapa* ssp. *pekinensis*, B) for one generation, no species-specific assortative mating was observed (PaB-PcB: I_{PSI} = 0.01 ± 0.08; $P = 0.95$). The males of both species copulated as long (PaB: $U = 557$, $P = 0.16$; PcB: $U = 617$, $P = 0.59$; Figure 3.1D) and as often (PaB: $\chi^2 = 0.72$, $P = 0.68$; PcB: $\chi^2 = 0.16$, $P = 1.00$; Figure 3.1E) with heterospecific females as with conspecific females.

Consistently, the males randomly attempted to mate with glass dummies treated with CHC extracts from either con- or heterospecific females (I_{PSI} = -0.05 ± 0.18, $P = 0.596$). They did not differ in their mating propensity towards these dummies (PaB: $\chi^2 = 0.05$, $P = 1.00$; PcB: $\chi^2 = 0.06$, $P = 1.00$; Figure 3.1F).

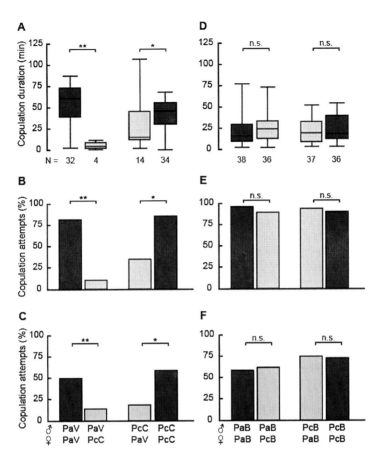

Figure 3.1: Impact of host plant use on interspecific behavioral isolation between two closely related mustard leaf beetle species, *Phaedon cochleariae* (Pc) and *P. armoraciae* (Pa). (*Left row*) Each beetle species was reared on its native host plant, i.e., either *Cardamine amara* (PcC) or *Veronica beccabunga* (PaV), respectively. (*Right row*) Both beetle species were reared on a common host, *Brassica rapa* ssp. *pekinensis*, (PcB and PaB). (*right row*). Mating durations (*A, D*) and male mating propensity towards live females fed on a respective host (*B, E*) and towards glass dummies treated with cuticular hydrocarbon extracts from these females (*C, F*) (*N* = 40 for each combination). Intra- and interspecific combinations are indicated by dark grey and light grey colors, respectively. The boxes represent 25% and 75% percentiles, the bars inside the boxes are the medians, and the whiskers represent the minima and maxima. The differences in mating duration and male mating propensity were analyzed by the Mann-Whitney *U* test and χ^2-test, respectively. *** P <0.001; ** P <0.01; * P <0.05; n.s. P >0.05.

The impact of host plant use on the beetles' cuticular hydrocarbon (CHC) phenotypes. In total, we detected 64 CHC peaks in the cuticular extracts of the four groups of beetles, i.e., *P. armoraciae* and *P. cochleariae* reared on either their native hosts or on Chinese cabbage (SI Table 3.1; SI Figure 3.2). The CHC profiles were complex mixtures of *n*-alkanes, mono- and dimethyl alkanes, unbranched olefins with one, two, or three double bonds, and monomethyl branched alkenes. All profiles were consistently dominated by 2-methylalkanes (PaV: 56.2±9.5%, PaB: 20.0±3.5%, PcC: 28.6±5.2%, PcB: 28.4±7.0%) and methyl-branched alkenes (PaV: 16.9±5.5%, PaB: 65.7±6.0 %, PcC: 45.3±9.2%, PcB: 47.8±7.7%).

A discriminant analysis based on the relative proportions of these 64 CHC peaks clearly separated the four groups of beetles (Wilks' λ < 0.001, $F_{192,693}$ = 161.23, P < 0.001; Figure 3.2). All individuals (N = 298, compare SI Table 3.1) were correctly classified by the original discriminant function, and 93.8% were cross-validated. The first two canonical roots accounted for 75.7% and 18.4% of the total variance and separated the CHC profile of *P. armoraciae* (PaV) reared on its native host from the remaining three groups. The CHC profile of *P. cochleariae* (PcC) reared on its native host was separated from the remaining three groups as well. The third canonical root, explaining 5.9% of the total variance, separated the CHC profiles of *P. armoraciae* (PaB) and *P. cochleariae* (PcB) reared on *B. rapa* ssp. *pekinensis*.

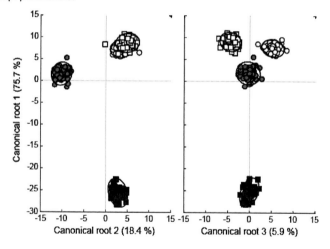

Figure 3.2: Discriminant analysis (DA) of cuticular hydrocarbon (CHC) phenotypes of *Phaedon cochleariae* and *P. armoraciae* in dependence of their host plants. The DA is based on CHC quantities that are listed in detail in SI Table 3.1. Scatterplot of canonical root 1 *vs.* 2 (A) and canonical root 1 *vs.* 3 (B) of the DA of CHC phenotypes of 21-d-old *P. cochleariae* feeding on *Cardamine amara* (grey circles; N = 80) or *Brassica rapa* ssp. *pekinensis* (white circles; N = 80) and of 21-d-old *P. armoraciae* reared on *Veronica beccabunga* (black squares; N = 58) or *B. rapa* ssp. *pekinensis* (white squares; N = 80), respectively. Variances explained by each root are given in parentheses.

The squared Mahalanobis distances of the group centroids of these CHC data sets reflected the results of the mating bioassays. When both species fed on their native host plant species, the chemical distance between their CHC profiles was much greater (PaV-PcC: 792.7; $F_{63,232}$ = 333.9, P < 0.001) than the distance between the profiles of the two species fed on the same host species (PaB-PcB: 90.8; $F_{63,232}$ = 45.5, P < 0.001). Moreover, the shift from brooklime (*V. beccabunga*) to Chinese cabbage (*B. rapa* spp. *pekinensis*) in *P. armoraciae* had a much stronger impact on the CHC profile (PaV-PaB: 1187.5; $F_{63,232}$ = 500.1, P < 0.001) than the host shift from large bittercress (*C. amara*) to Chinese cabbage in *P. cochleariae* (PcC-PcB: 358.5; $F_{63,232}$ = 179.6, P < 0.001).

3.5 Discussion

In spite of dietary resource partitioning, sympatric phytophagous insect species may frequently encounter each other when moving across a plant community while foraging for their host plants or mates. These insects depend on reliable behavioral isolation mechanisms that prevent reproductive interference with closely related species; otherwise, they are prone to sexual exclusion (Hochkirch et al. 2007; Gröning & Hochkirch 2008).

Our study showed that the two mustard leaf beetles *Phaedon armoraciae* and *P. cochleariae* rely on a phenotypically plastic isolation mechanism by producing mating signals, i.e., CHC phenotypes, depending on the host plant species they are feeding upon. Under natural conditions, the two mustard leaf beetle species co-occur in the same microhabitats but feed on different host plant species. We demonstrated that this dietary resource partitioning results in a divergence of their mate recognition systems. If the two species fed on different plant species, they exhibited distinct CHC phenotypes, which resulted in assortative mating with a clear preference for conspecific mates. However, when both beetle species fed upon the same plant species, their CHC phenotypes converged, and they did not discriminate anymore between con- and heterospecific mates because of this convergence of their mate recognition systems. Hence, the mechanism that impedes reproductive interference between the two phytophagous insect species is based on divergent host plant use acting in concert with the diet-induced phenotypic plasticity of the mate recognition systems. This mechanism impedes detrimental matings with heterospecific mates that do not result in any offspring due to unviable eggs (unpublished data).

The fundamentals of the biosynthesis of CHCs in insects are well known (Blomquist 2010). The majority of CHCs are synthesized *de novo* by a sophisticated network of fatty acid synthases, elongases, desaturases, and P450 oxidative decarbonylases (Blomquist 2010; Qiu et al. 2012; Chung et al. 2014). However, the exact mechanisms involved in the determination and regulation of CHC chain length or the position of methyl branches are poorly understood. Some evidence suggests that plant-derived compounds can interfere with the biosynthesis and regulation of CHCs in phytophagous insects (Blomquist & Jackson 1973; Etges et al. 2006; Otte et al. 2015). Moreover, Sharon et al. (2010) have demonstrated that a change in diet results in a change of the gut microbiome of *Drosophila melanogaster*, which in turn leads to altered CHC phenotypes and mate preferences. The composition of phyllosphere bacterial communities shows a considerable degree of species specificity and the tendency to follow host plant phylogeny (Kim et al. 2012; Vorholt 2012). Thus, the strong impact of a host shift between different plant families on the CHC phenotype in *P. armoraciae* (SI Table 3.1) may result from differences in plant nutrients and/or certain plant metabolites or from different bacterial communities on the leaves of Brassicaceae and Plantaginaceae.

A change of mating signals (CHC phenotypes) requires an adequate change of the preferences on the receiver side to guarantee the integrity of the mate recognition system. Although CHCs play a key role in insect communication (Singer 1998; Howard & Blomquist 2005), knowledge about the perception and processing of complex CHC profiles is still fragmentary (Ozaki et al. 2005). For many insects, the CHC phenotype matching mechanisms are believed to be involved in the chemosensory evaluation of encountered individuals (van Zweden & d'Ettorre 2010; Weddle et al. 2013). Geiselhardt et al. (2012) proposed that self-referent phenotype matching is the most likely mechanism to explain the coordinated shift of CHC phenotypes and mating preferences in *P. cochleariae* after a host plant switch. Self-referent phenotype matching implies that an individual uses its own phenotype as a template that is compared with the phenotype of an encountered individual (Dawkins 1982; Mateo 2004). The data of our present study support this suggestion. The similarity of CHC phenotypes of heterospecific beetles on a common host species was greater than that of conspecific beetles on different host plant species. Consequently, females feeding on the same host plant species as the males match the male phenotype more precisely than females from alternative hosts, irrespective of the beetle species. Consistently, beetles reared on the same host plant species mate randomly, whereas beetles reared on alternative host plant species showed host-specific assortative mating. The host-dependent mating preference does not only apply for heterospecific matings but also for conspecific matings of *P. armoraciae* (SI Figure 3.1) and *P. cochleariae* (Geiselhardt et al. 2012).

In self-referent phenotype matching, the signal of the sender and the response of the receiver are not considered to be two independent traits under stabilizing selection, but the plasticity of the signal and the response is expected to be based on the same principles (diet-induced plasticity of the CHC phenotype). However, self-referent phenotype matching only works if both sexes show similar reaction norms, i.e., if male and female phenotypes respond similar to the environmental conditions. Thus, stabilizing selection will act on the reaction norms of males and females rather than on a specific phenotype.

Whether the same mechanisms that cause intraspecific assortative mating are also involved in interspecific reproductive isolation is a long standing question. Most studies indicate that assortative mating within and between species do not share a common mechanism but are mediated by different (genetic) traits (Arbuthnott 2009; Bolnick & Kirkpatrick 2012; Laturney & Moehring 2012). In contrast, our results suggest that intraspecific mate recognition (SI Figure 3.1; Geiselhardt et al. 2012) and species recognition is based on the same (self-referent phenotype matching) mechanism. Host-induced phenotypic plasticity, assuming that it is sufficiently high to generate divergent phenotypes, can thus help to avoid maladaptive matings. Consequently, no genetic divergence of recognition systems is required to establish behavioral barriers to gene flow. Thus, this mechanism may not only play a substantial role in species coexistence but also in speciation.

Among evolutionary biologists, reproductive interference evokes great interest in terms of reinforcement and reproductive character displacement (Butlin 1987; Butlin 1989; Liou & Price 1994; Servedio & Noor 2003; Coyne & Orr 2004; Pfennig & Pfennig 2009; 2012). Reinforcement and reproductive character displacement both describe the process by which mate recognition systems diverge as a response to maladaptive interspecific reproductive interactions (Servedio & Noor 2003; Pfennig & Pfennig 2009). In both cases, the speciation process is initiated in allopatry, where geographically isolated populations accumulate genetic differences that cause a certain degree of postcopulatory reproductive isolation between them. However, host shifts are thought to be rare in allopatry (Tauber & Tauber 1989; Bush & Butlin 2011). Thus, after secondary contact, the diverged populations or species face sexual interference. The individuals who exhibit reproductive traits most dissimilar to those of heterospecific mates are favored by natural selection and gain a fitness benefit (Servedio & Noor 2003; Coyne & Orr 2004). Hence, according to these models, the behavioral isolation of diverging populations or species is based on genetic differences of the mate recognition system. In contrast, the behavioral isolation of the *Phaedon* species is based on host-dependent phenotypic traits that provide no target for divergent natural selection. Both resource competition and sexual interference may have forced one species to exploit an alternative host. This host shift achieves two goals at once: first, the avoidance of resource competition and, second, behavioral isolation mediated by host-induced divergence of CHC phenotypes.

Alternatively, speciation may start in sympatry with the evolution of divergent host preferences (Diehl & Bush 1989). However, alternative host plant use is not necessarily linked with spatial isolation; thus, cross-mating between populations may occur that counteracts local adaptation and prevents population divergence. In the case of *P. cochleariae* and *P. armoraciae*, alternative host use leads immediately to divergent host–associated CHC phenotypes and behavioral isolation between these CHC phenotypes, which can promote local adaptations and further genetic divergence (Fitzpatrick 2012; Geiselhardt et al. 2012). Thus, our results highlight the potential role of phenotypic plasticity of mate recognition systems as a barrier to gene flow both during incipient speciation and between closely related species in sympatry.

In conclusion, divergent host use of closely related species can induce the phenotypic divergence of mate recognition systems. Thus, alternative host use can not only lead to spatial and temporal isolation (Coyne & Orr 2004) but also to behavioral isolation. The phenotypic plasticity of mate recognition systems can eliminate the selection pressure that otherwise may drive the evolution of genetically based divergences of mate recognition systems between diverging populations or species. Consequently, two phytophagous insect species can coexist without genetically diverged mate recognition systems if they differ in host plant use. However, the efficiency of phenotypic plasticity as a barrier to gene flow relies on both the impact of alternative hosts on the phenotype and the host fidelity. Nevertheless, our data indicate that genetically driven divergence in mate recognition systems is no prerequisite for the avoidance of sexual interference in closely related species.

3.6 Supplemental Data

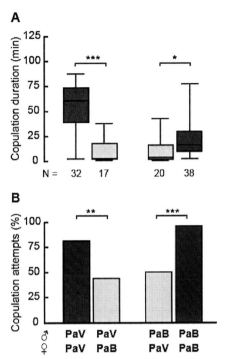

SI Figure 3.1: Impact of alternative host plant use on intraspecific behavioral isolation of the mustard leaf beetle *Phaedon armoraciae* (Pa). Mating durations (A) and mating propensities (B) of "same host plant" couples (dark grey columns) and "different host plant" couples (light grey columns). The beetles were reared on their native host plant species *Veronica beccabunga* (V) or on *Brassica rapa* ssp. *pekinensis* (B) (*N* = 40 for each combination). The boxes represent 25% and 75% percentiles, the bars inside the boxes are the medians, and the whiskers represent the minima and maxima. Differences in mating duration and male mating propensity were analyzed by the Mann-Whitney *U* test and χ^2-test, respectively. *** P <0.001; ** P <0.01; * P <0.05.

SI Table 3.1: Mean (± S.D.) relative composition (% total peak area) of cuticular hydrocarbons of *Phaedon cochleariae* and *P. armoraciae* reared on their native host plant *Cardamine amara* (PcC) and *Veronica beccabunga* (Pav), respectively, or on *Brassica rapa* ssp. *pekinensis*, (PcB and PaB).

Retention index	Compound	PcB (N=80)	PcC (N=80)	PaB (N=80)	Pav (N=58)
1943	7-MeC19	0.2±0.2	0.4±0.4	-	-
2269	6,9-C23diene	2.9±1.8	2.9±1.7	0.3±0.3	0.5±1.0
2274	9-C23ene	1.7±1.0	0.9±0.5	0.3±0.3	1.5±1.7
2462	2-MeC24	2.5±1.3	4.5±1.9	2.7±2.1	0.5±0.3
2476	6,9-C25diene	0.0±0.1	0.7±0.5	0.2±0.2	-
2479	3,6,9-C25triene	0.6±0.6	0.8±0.5	0.8±0.5	-
2482	7-C25ene	0.2±0.1	0.4±0.3	0.2±0.1	-
2500	n-C25	0.1±0.1	0.4±0.2	0.2±0.2	0.8±0.4
2507	13-/15-MeC25-7-ene	0.3±0.2	0.4±0.2	0.2±0.1	-
2534	11-/13-MeC25	0.9±0.5	1.9±0.9	1.3±0.9	-
2563	2-MeC25	0.1±0.1	0.2±0.1	-	0.3±0.2
2573	7,13-dimeC25	0.5±0.2	0.6±0.2	0.2±0.1	0.3±0.2
2582	5,13-dimeC25	0.1±0.1	0.2±0.9	-	0.4±0.3
2609	14-MeC26-3/8-ene^a / 15-MeC26-9-ene	0.2±0.1	0.4±0.3	0.2±0.2	0.4±0.3
2664	2-MeC26	4.3±1.2	4.7±1.5	1.6±0.6	16.6±4.1
2676	6,9-C27diene	0.3±0.2	1.1±0.8	0.3±0.2	-
2676	9-C27ene	0.2±0.4	0.5±0.5	0.3±0.6	2.7±1.0
2682	3,6,9-C27triene	1.1±1.2	1.0±1.1	1.4±1.2	-
2707	13-MeC27-2-ene / 15-MeC27-4/9-ene	2.0±0.9	3.6±1.6	1.6±0.7	7.3±3.4
2732	13-MeC27	1.4±0.7	1.3±0.5	0.9±1.0	0.3±0.3
2764	2-MeC27 / 9,13-/9,15-dimeC27	0.6±0.1	0.6±0.2	0.3±0.2	0.7±0.2
2782	5,13-/5,15-dimeC27	1.3±0.5	1.1±0.5	0.4±0.2	0.3±0.2
2800	n-C28	0.4±0.3	0.2±0.2	-	0.9±0.3
2834	12-/13-/14-MeC28	-	0.2±0.4	-	-
2862	2-MeC28	17.0±6.1	14.2±3.3	9.7±2.1	25.6±6.0
2876	6,9-C29diene	0.3±0.4	0.6±0.8	-	-
2878	9-C29ene	0.2±0.3	0.1±0.2	-	2.9±2.0
2882	3,6,9-C29triene	0.1±0.1	0.2±0.3	-	-
2900	n-C29	0.3±0.1	0.4±0.1	0.2±0.1	2.8±0.7
2931	13-/15-MeC29	1.1±0.4	1.0±0.5	0.3±0.1	0.4±0.5
2963	2-MeC29	0.2±0.1	0.1±0.1	0.2±0.1	0.4±0.2
2982	5,13-/5,15-dimeC29	-	0.1±0.1	-	0.4±0.3
3005	3,13-/3,15-dimeC29	0.2±0.1	-	-	-
3063	2-MeC30	4.1±2.4	4.0±1.4	5.6±2.1	12.8±6.4
3084	9-C31ene	-	-	0.4±0.8	3.3±2.0
3102	2,12-/2,14-dimeC30	0.1±0.2	1.2±1.4	-	-
3109	17-MeC31-6/8-ene / 19-MeC31-8/10-ene	0.4±0.3	0.6±0.6	-	-
3132	13-MeC31	1.2±0.4	1.4±0.7	0.3±0.3	0.1±0.3
3205	3,13-/3,15-dimeC31	0.5±0.2	1.1±0.9	0.2±0.3	-
3261	2-MeC32	-	0.7±1.3	0.2±0.2	-
3281	5,13-dimeC32	-	-	0.4±0.4	-
3331	13-MeC33	0.8±0.2	0.6±0.4	0.4±0.1	2.8±1.9
3382	5,13-dimeC33	1.4±0.4	1.9±0.8	0.9±0.4	2.6±0.9
3405	3,13-dimeC33	0.3±0.1	0.2±0.2	0.1±0.1	0.4±0.4
3506	21-MeC35-10/12-ene / 23-MeC35-12/14-ene	4.6±2.4	2.3±2.2	0.4±0.6	-
3532	13-/15-MeC35	0.8±0.2	0.5±0.3	0.7±0.2	-
3572	7,13-/7,15-dimeC35	0.4±0.2	-	0.2±0.2	-
3582	5,13-/5,15-dimeC35	1.2±0.3	0.2±0.2	1.5±0.4	0.2±0.2
3607	23-MeC36-12-ene / 24-MeC36-13-ene / 25-MeC36-14-ene	0.5±0.4	1.0±0.4	0.6±0.3	2.5±1.0
3605	3,13-/3,15-dimeC35	0.4±0.4	0.5±0.4	0.0±0.0	-
3708	23-MeC37-12/14-ene / 25-MeC37-14/16-ene	6.3±1.7	7.4±2.7	7.9±1.6	-
3731	13-/15-MeC37	0.3±0.2	-	0.9±0.5	-
3772	7,13-/7,15-dimeC37	0.3±0.4	-	-	-
3808	25-MeC38-14-ene / 26-MeC38-15-ene / 27-MeC38-16-ene	0.1±0.2	0.9±0.6	1.2±0.4	0.8±0.9
3804	3,13-/3,15-dimeC37	0.2±0.2	-	-	-
3909	25-MeC39-14/16-ene / 27-MeC39-16/18-ene	10.4±5.7	14.8±5.2	28.0±3.4	3.8±2.2
3933	13-/15-MeC39	-	-	0.3±0.4	-
4008	27-MeC40-14/16-ene / 28-MeC40-17-ene / 29-MeC40-18-ene	0.4±0.3	0.6±0.6	1.0±0.4	-
*	27-MeC41-16/18-ene / 29-MeC41-18/20-ene	7.3±2.7	6.6±2.1	12.8±1.7	0.3±0.5
*	29-MeC42-18-ene / 30-MeC42-19-ene / 31-MeC42-20-ene	0.7±0.3	0.5±1.1	0.8±0.5	1.1±1.0
*	29-MeC43-18/20-ene / 34-MeC43-20/22-ene	12.6±3.5	5.5±2.3	8.8±2.6	0.3±0.8
*	32-MeC44-21-en / 31-MeC44-20-en / 33-MeC44-22-ene	0.4±0.3	0.9±1.2	-	0.3±0.8
*	31-MeC45-20/22-ene / 33-MeC45-22/24-ene	1.7±0.6	-	2.1±0.7	-

^a 14-MeC26-3/8-ene refers to 14-MeC26-3-ene + 14-MeC26-8-ene.* No reference alkanes were available

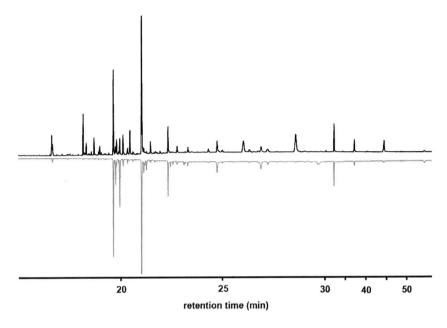

retention time (min)

SI Figure 3.2: Total ion current chromatograms of the cuticular hydrocarbons extracted with dichloromethane from female *Phaedon cochleariae* (black), feeding on large bittercress (*Cardamine amara*), and of female *P. armoraciae* (grey), reared on brooklime (*Veronica beccabunga*).

3.7 References

Aitchison J. (1986). The Statistical Analysis of Compositional Data. *Chapman and Hall, London.*

Arbuthnott D. (2009). The genetic architecture of insect courtship behaviour and premating isolation. *Heredity*, 103, 15-22.

Berlocher S.H. & Feder J.L. (2002). Sympatric speciation in phytophagous insects: Moving beyond controversy? *Annual Review of Entomology*, 47, 773-815.

Blomquist G.J. (2010). Biosynthesis of cuticular hydrocarbons, pp. 35-52, in G.J. Blomquist and A.G. Bagnères (eds.). Insect Hydrocarbons: Biology, Biochemistry and Chemical Ecology. *Cambridge University Press, Cambridge.*

Blomquist G.J. & Jackson L.L. (1973). Incorporation of labelled dietary *n*-alkanes into cuticular lipids of the grasshopper *Melanoplus sanguinipes*. *Journal of Insect Physiology*, 19, 1639-1647.

Böhme J. (2001). Phytophage Käfer und ihre Wirtspflanzen in Mitteleuropa - Ein Kompendium. *Bioform, Heroldsberg.*

Bolnick D.I. & Kirkpatrick M. (2012). The relationship between intraspecific assortative mating and reproductive isolation between divergent populations. *Current Zoology*, 58, 484-492.

Bush G.L. & Butlin R.K. (2011). Sympatric speciation in insects, pp. 229-248, in U. Diekmann, M Doebli, J.A.J. Metz und D. Tautz (eds.). Adaptive Speciation. *Cambridge University Press, Cambridge.*

Butlin R.K. (1987). Speciation by reinforcement. *Trends in Ecology & Evolution*, 2, 8-13.

Butlin R.K. (1989). Reinforcement of premating isolation, pp. 158-179, in D. Otte and J.A. Endler (eds.). Speciation and its Consequences. *Sinauer, Sunderland.*

Butlin R.K. & Ritchie M.G. (1994). Mating behaviour and speciation, pp. 43-79, in P.J.B. Slater and T.R. Halliday (eds.). Behaviour and Evolution. *Cambridge University Press, Cambridge.*

Carvajal-Rodríguez A. & Rolán-Alvarez E. (2006). JMATING: a software for the analysis of sexual selection and sexual isolation effects from mating frequency data. *BMC Evolutionary Biology*, 6, 40.

Chung H., Loehlin D.W., Dufour H.D., Vaccarro K., Millar J.G. & Caroll S.B. (2014). A single gene affects both ecological divergence and mate choice in *Drosophila. Science*, 343, 1148-1151.

Cornwallis C.K. & Uller T. (2010). Towards an evolutionary ecology of sexual traits. *Trends in Ecology & Evolution*, 25, 145-152.

Coyne J.A. (1992). Genetics and speciation. *Nature*, 355, 511-515.

Coyne J.A. & Orr H.A. (2004). Speciation. *Sinauer Associates, Sunderland.*

Dawkins R. (1982). The Extended Phenotype. *W.H. Freeman, San Francisco.*

Denno R.F., McClure M.S. & Ott J.R. (1995). Interspecific interactions in phytophagous insects: competition reexamined and resurrected. *Annual Review of Entomology*, 40, 297-331.

Diehl S.R. & Bush G.L. (1989). The role of habitat preference in adaptation and speciation, pp. 345-365, in D. Otte and J.A. Endler (eds.). Speciation and its Consequences. *Sinauer, Sunderland.*

Edwards C.A. & Heath G.W. (1964). The Principles of Agricultural Entomology. *Chapman & Hall, London.*

Etges W.J. (1992). Premating isolation is determined by larval rearing substrates in cactophilic *Drosophila mojavensis. Evolution*, 46, 1945-1950.

Etges W.J., Veenstra C.L. & Jackson L.L. (2006). Premating isolation is determined by larval rearing substrates in cactophilic *Drosophila mojavensis*. VII. Effects of larval dietary fatty acids on adult epicuticular hydrocarbons. *Journal of Chemical Ecology*, 32, 2629-2646.

Fitzpatrick B.M. (2012). Underappreciated consequences of phenotypic plasticity for ecological speciation. *International Journal of Ecology*, 2012, Article ID 256017.

Funk D.J., Filchak K.E. & Feder J.L. (2002). Herbivorous insects: model systems for the comparative study of speciation ecology. *Genetica*, 116, 251-267.

Futuyma D.J. & Peterson S.C. (1985). Genetic variation in the use of resources by insects. *Annual Review of Entomology*, 30, 217-238.

Geiselhardt S., Otte T. & Hilker M. (2009). The role of cuticular hydrocarbons in male mating behavior of the mustard leaf beetle, *Phaedon cochleariae* (F.). *Journal of Chemical Ecology*, 35, 1162-1171.

Geiselhardt S., Otte T. & Hilker M. (2012). Looking for a similar partner: host plants shape mating preferences of herbivorous insects by altering their contact pheromones. *Ecology Letters*, 15, 971-977.

Green M.D. (2002). Signalers and Receivers. *Oxford University Press, New York.*

Gröning J. & Hochkirch A. (2008). Reproductive interference between animal species. *The Quarterly Review of Biology*, 83, 257-282.

Hardin G. (1960). The competitive exclusion principle. *Science*, 131, 1292-1297.

Hochkirch A., Gröning J. & Bücker A. (2007). Sympatry with the devil – Reproductive interference could hamper species coexistence. *Journal of Animal Ecology*, 76, 633-642.

Howard R. & Blomquist G.J. (2005). Ecological, behavioral, and biochemical aspects of insect hydrocarbons. *Annual Review of Entomology*, 50, 371-393.

Kaplan I. & Denno R.F. (2007). Interspecific interactions in phytophagous insects revisited: a quantitative assessment of competition theory. *Ecology Letters*, 10, 977-994.

Kim M, et al. (2012). Distinctive phyllosphere bacterial communities in tropical trees. *Microbial Ecology*, 63, 674-681.

Kuno E. (1992) Competitive exclusion through reproductive interference. *Research in Population Ecology*, 34, 275-284.

Laturney M. & Moehring A.J. (2012). The genetic basis of female mate preference and species isolation in *Drosophila*. *International Journal of Ecology*, 2012, Article ID 328392.

Liou L.W. & Price T.D. (1994). Speciation by reinforcement of premating isolation. *Evolution*, 48, 1451-1459.

Mateo J.M. (2004). Recognition systems and biological organization: The perception component of social recognition. *Annales Zoologici Fennici*, 41, 729-745.

Matsubayashi K.W., Ohshima I. & Nosil P. (2010) Ecological speciation in phytophagous insects. *Entomologia Experimentalis et Applicata*, 134, 1-27.

Nosil P. (2012). Ecological speciation. *Oxford University Press, Oxford.*

Otte T., Hilker M. & Geiselhardt S. (2015). The effect of dietary fatty acids on the cuticular hydrocarbon phenotype of an herbivorous insect and consequences for mate recognition. *Journal of Chemical Ecology*, 41, 32-43.

Ozaki M., Wada-Katsumata A., Fujikawa K., Iwasaki M., Yokohari F., Satoji Y., Nisimura T. & Yamaoka R. (2005). Ant nestmate and non-nestmate discrimination by a chemosensory sensillum. *Science*, 309, 311-314.

Pfennig D.W. & Pfennig, K.S. (2012). Development and evolution of character displacement. *Annals of the N.Y. Academy of Sciences*, 1256, 89-107.

Pfennig K.S. & Pfennig D.W. (2009). Character displacement: ecological and reproductive responses to a common evolutionary problem. *The Quarterly Review of Biology*, 84, 253-276.

Qiu Y, et al. (2012). An insect-specific P450 oxidative decarbonylase for cuticular hydrocarbon biosynthesis. *Proceedings of the National Academy of Sciences of the United States of America*, 109, 14858-14863.

Reitz S.R. & Trumble J.T. (2002). Competitive displacement among insects and arachnids. *Annual Review of Entomology*, 47, 435-465.

Ribeiro J.M.C. & Spielman A. (1986). The satyr effect: a model predicting parapatry and species extinction. *American Naturalist*, 128, 513-528.

Rolán-Alvarez E. & Caballero A. (2000). Estimating sexual selection and sexual isolation effects from mating frequencies. *Evolution*, 54, 30-36.

Rundle H.D. & Nosil P. (2005). Ecological speciation. *Ecology Letters*, 8, 336-352.

Schoener T.W. (1974). Resource partitioning in ecological communities. *Science*, 185, 27-39.

Servedio M.R., Van Doorn G.S., Kopp M., Frame A.M. & Nosil, P. (2011) Magic traits in speciation: 'magic' but not rare? *Trends in Ecology & Evolution*, 26, 389-397.

Sharon G., Segal D., Ringo J.M., Hefez A., Zilber-Rosenberg I. & Rosenberg E. (2010). Commensal bacteria play a role in mating preference of *Drosophila melanogaster*. *Proceedings of the National Academy of Sciences of the United States of America*, 107, 20051-20056.

Singer T. (1998). Roles of hydrocarbons in the recognition systems of insects. *American Zoologist*, 38, 394-405.

Strong D.R., Lawton J.H. & Southwood T.R.E. (1984). Insects on Plants. *Harvard University Press, Cambridge.*

Tauber C. & Tauber M.J. (1989). Sympatric speciation in insects: perception and perspective, pp. 307-344, in D. Otte and J.A. Endler (eds.). Speciation and Its Consequences. *Sinauer, Sunderland.*

Van Zweden J.S. & d'Ettorre P. (2010). Nestmate recognition in social insects and the role of hydrocarbons, pp 222–243, in G.J. Blomquist and A.G. Bagnères (eds.). Insect Hydrocarbons: Biology, Biochemistry and Chemical Ecology. *Cambridge University Press, Cambridge.*

Vorholt J.A. (2012). Microbial life in the phyllosphere. *Nature Reviews Microbiology*, 10, 828-840.

Webster S.E., Galindo J., Grahame J.W. & Butlin R.K. (2012). Habitat choice and speciation. *International Journal of Ecology*, 2012, Article ID 154686.

Weddle C.B., Hunt J. & Sakaluk S.K. (2013). Self-referent phenotype matching and its role in female mate choice in arthropods. *Current Zoology*, 59, 239-248.

The Effect of Dietary Fatty Acids on the Cuticular Hydrocarbon Phenotype of an Herbivorous Insect and Consequences for Mate Recognition

4.1 Abstract

The cuticular hydrocarbon (CHC) profile of the mustard leaf beetle *Phaedon cochleariae* is known to serve mate recognition and to depend on the food plant species; beetles were previously shown to prefer mates that are feeding on the same plant species and have a similar beetle CHC pattern. In order to elucidate whether the pattern of ingested fatty acids that may be used for CHC biosynthesis affects the CHC pattern of *P. cochleariae* adults, we fed the beetles (a) with two different host plant species differing in their fatty acid profiles and (b) with artificial diets differing mainly in their compositions of mono-, di-, and triunsaturated fatty acids. The analyses of the beetles´ CHC patterns revealed that ingestion of different fatty acid blends results in significant effects on the beetle´s quantitative pattern of straight-chained and methyl-branched CHCs. Interestingly, the CHC patterns of males and females were differently affected by ingestion of fatty acids. In contrast to beetles feeding on different host plant species, behavioral bioassays revealed that mating preferences of beetles fed with different types of artificial diet were not affected by their diet-dependent CHC profiles. We suggest that the occurrence of CHC-dependent assortative mating in *P. cochleariae* does not depend on the dietary fatty acids offered to the beetles in this study, but on other food constituents that affect their CHC biosynthesis.

4.2 Introduction

Ecological speciation is a process in which ecologically driven disruptive selection leads to evolution of reproductive isolation between populations in divergent environments (Schluter 2001; Rundle & Nosil 2005; Schluter 2009; Nosil 2012).

In herbivorous insects that are highly specialized on one or few plant species, a switch to a novel host species represents a strong divergent selection pressure that favors local adaptations and thus may initiate ecological speciation (Egan & Funk 2009; Via 2009). Local adaptation to alternative host species may lead to natural selection against migrants or hybrids (Nosil et al. 2005; Egan & Funk 2009; Via 2009) and/or to host-specific assortative mating due to habitat or temporal isolation (Coyne & Orr 2004; Matsubayashi et al. 2010). Moreover, ecological parameters can affect sexual communication and thus interact with sexual selection in the evolution of behavioral isolation (Maan & Seehausen 2011; Wilkins et al. 2013; Safran et al. 2013). For example, transmission and perception of visual and acoustical mating signals may depend on the environment, and thus diverging environments may drive behavioral isolation (Boughman 2001; 2002). Biosynthesis and release of chemical mating signals (pheromones) may depend on the host plant species of herbivorous insects and thus, may directly impact on behavioral isolation if colonization of alternative host plants generates insect phenotypes that differ in traits involved in sexual communication (Smadja & Butlin 2009; Fitzpatrick 2012; Geiselhardt et al. 2012; The Marie Curie SPECIATION Network 2012).

Insect cuticular hydrocarbons (CHCs) are known to serve as contact pheromones and to play a crucial role in mate recognition in several insect species (Singer 1998; Howard & Blomquist 2005). When the pattern of these CHC-based mate recognition pheromones depends on the nutritional resource, CHC profile alteration resulting from a change in the nutritional resource may lead to behavioral isolation. Indeed, Sharon et al. (2010) have shown that commensal bacteria associated with the food media alter the CHC phenotype of *Drosophila melanogaster* flies and thus cause diet-specific assortative mating. In the cactophilic *D. mojavensis*, the larval rearing substrate affects both adult CHC phenotypes and mating behavior (Etges 1992; Brazner & Etges 1993). Etges et al. (2006) have shown that the quantitative and qualitative composition of a blend of triacylglycerides in the larval diet have a significant effect on CHC profiles of adult *D. mojavensis*.

Food-dependence of insect CHC profiles was not only shown in *Drosophila* species, but also in parasitoid wasps (Kühbander et al. 2012) and the mustard leaf beetle *Phaedon cochleariae* that use their CHCs for mate recognition (Geiselhardt et al. 2009; 2012). This leaf beetle occurs in most parts of Europe and spreading eastwards to Sibiria. The host plant range of *P. cochleariae* includes several members of the tribe Cardamineae (Brassicaceae), mainly watercress (*Nasturtium officinale*) and large bittercress (*Cardamine amara*), common species growing on banks along brooks and in marshes. A shift to a novel host plant species alters the CHC phenotype of adult beetles within two weeks. This plant-induced divergence in CHC phenotypes of beetles feeding on different plant species causes host-specific assortative mating (Geiselhardt et al. 2012).

Ingested plant material may affect the beetle´s CHC pattern both by providing CHC precursors and by affecting enzyme activity involved in de novo CHC biosynthesis by the beetles. Most insects synthesize their CHCs de novo by elongation of fatty acid acyl-CoAs to very long-chained fatty acids and subsequent decarboxylation to hydrocarbons (Blailock et al. 1976; Howard & Blomquist 2005; Blomquist 2010). However, insects may also use plant-derived dietary fatty acids as precursors for the biosynthesis of pheromonal hydrocarbons by elongation and decarboxylation of the fatty acids (Stanley-Samuelson et al. 1988; Pennanec'h et al. 1997; Millar 2000; 2010). Hence, the qualitative and quantitative blend of plant fatty acids might affect insect CHC profiles.

In the present study, we addressed the question whether the composition of dietary fatty acids ingested by *P. cochleariae* affects the CHC phenotype of adults, and thus mating preferences. The CHC profiles of *P. cochleariae* are complex mixtures composed of *n*-alkanes, methyl-branched alkanes, methyl-branched alkenes, and straight-chained olefins with one, two, or three double bonds (Geiselhardt et al. 2009; 2012). Double bonds in straight-chained olefins are located at positions 7 and 9 in monoenes, 6,9 in dienes, and 3,6,9 in trienes. Thus, these hydrocarbons might be derived from dietary fatty acid precursors, e.g. vaccenic acid (C18:1,n-7), oleic acid (C18:1,n-9), linoleic acid (C18:2,n-6), and linolenic acid (C18:3,n-3). Furthermore, saturated fatty acids can serve as precursors for *n*-alkanes. Previous studies have shown that the alkene fraction of *P. cochleariae* CHCs per se does not elicit any mating behavior, whereas the alkane fraction or a mixture of alkanes and alkenes can elicit male copulation behavior (Geiselhardt et al. 2009). Here, we focused on the relevance of uptake of dietary fatty acids on the beetle´s CHC profile and their impact on mating behavior.

In a first step, we analyzed the fatty acid composition of two host plant species, i.e. Chinese cabbage (*Brassica rapa* ssp. *pekinensis*) and watercress (*N. officinale*), and investigated how the plant´s fatty acid composition affects the CHC phenotype of the beetles. To scrutinize the impact of dietary fatty acids on the CHC profiles, we reared beetles on three alternative semi-artificial diets differing mainly in the compositions of fatty acids and analyzed the beetles´ CHC pattern. To test whether divergences in CHC phenotypes induced by feeding on the different artificial diets resulted in diet-specific assortative mating, we conducted mating bioassays with all possible male × female combinations.

4.3 Materials and Methods

Beetles and Plants. All *Phaedon cochleariae* beetles originated from a laboratory stock reared for multiple generations on Chinese cabbage in a climate chamber at 20°C, 70% relative humidity, and a 16:8 h light-dark cycle.

In order to investigate the impact of divergent host plant use on the CHC pattern of *P. cochleariae* adults, we reared beetles for one generation on two different brassicaceous host plant species, i.e. Chinese cabbage (*Brassica rapa* ssp. *pekinensis*) and watercress (*Nasturtium officinale*). Both host plants were grown in a greenhouse. Newly hatched larvae from the stock culture on Chinese cabbage were placed in a plastic box (20 x 20 x 6.5 cm; Gerda, Schwelm, Germany) that was covered with gauze and lined with moist paper towels. Larvae were supplied with fresh leaves of either host plant species *ad libitum* until the pupal stage. Emerging adults were separated by sex and supplied with fresh leaves of either host plant species. We used virgin 42-day-old, plant-fed adult beetles for the analyses of their CHC patterns.

Chinese cabbage and watercress seeds were obtained from Saatzucht Quedlinburg GmbH (Quedlinburg, Germany) and Carl Sperling & Co. GmbH (Lüneburg, Germany), respectively. Plants were grown in soil in a greenhouse under long day conditions (photoperiod 16 h, photosynthetic photon flux 175 µmol m^{-2} sec^{-1}, temperature 22±4°C, relative humidity 35±10%). Leaves from non-flowering 4-week-old plants were used for extraction of leaf fatty acids and as beetle food.

To elucidate the impact of artificial diets that mainly differ in their fatty acid composition on CHC profiles of adult *P. cochleariae*, we transferred pupae from the stock population (fed with Chinese cabbage during larval development) individually into plastic Petri dishes (Ø 5.5 cm), lined with moist filter paper. After eclosion, the adult beetles were supplied with a freshly prepared piece of semi-synthetic artificial diet (approximately 0.5 g; see below) every 3 d. Survival rates and weight of adult *P. cochleariae* did not differ between the three types of artificial diet (SI Table 4.1). Adults fed with the artificial diets described here showed survival rates between 21-41%, whereas survival rates of adults fed with Chinese cabbage ranged between about 76-90%. In contrast to survival rates, weight of adults fed with artificial diets hardly differed from weight of adults fed with Chinese cabbage (SI Table 4.1). We used virgin 42-day-old, diet-fed adults for CHC analyses and for the bioassays.

Artificial Diets. Beetles were offered three different diet mixtures varying mainly in their composition of fatty acids. For all diets, 20 ml deionized water was blended with wheat germ (2.5 g) (MP Biomedicals, Illkirch, France), casein (2 g) (Sigma Aldrich, Steinheim, Germany), salt mixture W (0.5 g) (MP Biomedicals), sorbic acid (50 mg) (Carl Roth, Karlsruhe; Germany), ascorbic acid (250 mg) (Carl Roth), cholesterol (250 mg) (Sigma Aldrich), and agar (2 g) (Carl Roth). This mixture was boiled on a heater while stirring constantly at low speed. After the blend had cooled down to 60 °C, we added a Vanderzant vitamin mixture (500 mg), methyl paraben (50 mg), β-sitosterol (50 mg) (all Sigma Aldrich), lyophilized Chinese cabbage leaf powder (1.5 g), vegetable oil (250 µl) (commercial brands, see below for suppliers), and 10 ml deionized water. The entire mixture was blended for several minutes, poured into a glass Petri dish, allowed to cool down, and then exposed to UV light for 20 min for sterilization. Thereafter, each Petri dish was covered with parafilm and stored at -20°C until use.

The three diets differed only in the type of vegetable oil added. We used vegetable oils that vary substantially in their fatty acid composition and chose olive oil (Primadonna, Lidl, Neckarsulm, Germany) with a high concentration of monounsaturated fatty acids (high Cx:1), sunflower oil (A&P, Kaisers Tengelmann, Viersen, Germany) with a high concentration of di-unsaturated fatty acids (Cx:2), and linseed oil (Kunella Feinkost, Cottbus, Germany) with high concentration of fatty acids with three double bonds. In detail, the fatty acid composition of the oils was: olive oil (Cx:0 = 16.2%; Cx:1 = 72.2%; Cx:2 = 10.4%; Cx:3 = 0.9%), sunflower oil (Cx:0 = 10.6%; Cx:1 = 24.8%; Cx:2 = 63.9%; Cx:3 = 0.7%), and linseed oil (Cx:0 = 8.4%; Cx:1 = 18.6%; Cx:2 = 14.1%; Cx:3 = 58.9%). The fatty acid compositions of the vegetable oils (10 µl oil added to 90 µl n-hexane) were determined by transesterification to fatty acid methyl ester (FAME) derivatives as described below (section *Extraction of Fatty Acids from Host Plant Leaves*).

Extraction of Cuticular Hydrocarbons from Beetles. Prior to extraction, beetles were killed by freezing at -20°C. Crude extracts were prepared by immersing individual beetles for 10 min in 150 µl dichloromethane (DCM). These extracts were concentrated to dryness under a gentle nitrogen stream, re-dissolved in 50 µl DCM, and stored at -20°C until use. Extracts of 40 individual beetles (20 males and 20 females) per host plant species and type of artificial diet, respectively, were prepared for chemical analysis.

Extraction of Fatty Acids from Host Plant Leaves. To analyze the fatty acid composition of the host plants, fresh leaf tissue (1 cm²) of each plant species was treated with 0.5 ml methanolic-HCl (450 µl MeOH + 50 µl HCl) at 70°C for 1 h to transesterify the lipids into FAMEs. After cooling to room temperature, 0.5 ml hexane (containing 1 µg n-eicosane as internal standard) was added in order to extract the FAMEs (Browse et al. 1986). Prior to derivatization, we determined the fresh weight of each piece of leaf. Ten extracts (replicates) per plant species were prepared and stored at -20°C until analysis.

Chemical Analysis. All (plant and insect) extracts were analyzed by coupled gas chromatograph-mass spectrometry (GC-MS) (7890 GC - 5975C MSD; Agilent Technologies, Waldbronn, Germany). Our GC-MS system was equipped with a programmable temperature injection unit (CIS-4; Gerstel, Mühlheim a. d. Ruhr, Germany). Electron impact ionization was 70 eV. The system was operated with a constant helium flow of 1 ml/min.

Samples (1 µl) of extracts from beetles were injected at 150°C in splitless mode. The CIS-4 was immediately heated to 300°C at 12°C/sec. The oven was programmed from 100°C to 320°C (35 min isotherm) at a rate of 10°C/min.

Samples were separated on a fused silica column (DB-5MS, 30 m x 0.32 mm ID x 0.25 µm, J & W Scientific, Folsom, USA).

Samples (1 µl) of FAME extracts obtained from leaf tissue were injected at 50°C, and the CIS-4 was immediately heated to 280°C at 12°C/sec. A DB-wax column (30 m x 0.25 mm ID x 0.25 µm, J & W Scientific, Folsom, CA, USA) was used. The oven temperature increased from initially 40°C to 240°C (10 min isotherm) at a rate of 10°C/min.

Identification of the beetles' CHCs was performed as previously described by Geiselhardt et al. (2009). Plant lipids and FAMEs were identified by comparison of their fragmentation patterns with those in the Wiley mass spectra library (John Wiley & Sons, Ltd.), and by comparison of their mass spectra and retention times with those of authentic standards. All reference compounds were purchased from Sigma-Aldrich (Munich, Germany). Double-bond positions were determined after derivatization with dimethyl disulfide (Francis & Veland 1981; Scribe et al. 1988). The double-bond positions of the 3,6,9-trienes were determined by their characteristic mass fragments, i.e. m/z 79, 108, and a loss of m/z 56 from the molecular ion (Millar 2000).

Quantification of fatty acids, n-alkanes, and methyl-branched hydrocarbons in the extracts was based on peak areas obtained from total ion current chromatograms. Because of overlapping peaks, peak areas of straight-chained unsaturated hydrocarbons were obtained from single-ion chromatograms (molecular mass of the respective compound). Relative quantities of compounds were determined by calculating peak areas of the compounds of interest relative to total peak areas of total ion chromatograms; these relative quantities were used for statistical comparisons.

Behavioral Bioassays. To study *P. cochleariae* mating preferences in dependence of the dietary fatty acid composition ingested by the mating partners, we tested all possible couple combinations of beetles (42-d-old virgin beetles) reared on either type of artificial diet. Mating bioassays were conducted as described by Geiselhardt et al. (2009). Bioassays were started by placing a test male into a Petri dish (5.5 cm x 1.2 cm) containing a potential mate. We recorded whether a test male started a copulation attempt, and if so, how long this attempt lasted. Each beetle was tested only once. We tested 40 couples of each possible combination. Mating preferences of beetles feeding on Chinese cabbage or watercress had been tested in a previous study and had shown that males preferred females feeding on the same host species to females feeding on alternative host species (Geiselhardt et al. 2012).

Statistical Analysis. All statistical analyses were conducted using SPSS 18 (SPSS Inc., Chicago, IL, USA) or Statistica 7.0 (StatSoft Inc., Tulsa, OK, USA).

The fatty acid patterns of both host plant species were compared by a MANOVA. Differences in the total fatty acid content (based on tissue fresh weight) and in relative quantities of single fatty acids between plant species were statistically analyzed by the Mann-Whitney U test. We only detected a significant difference between plant species when grouping (i) all monounsaturated and (ii) all diunsaturated fatty acids. Thus, we compared the ratios of mono- and diunsaturated fatty acids between the two plant species and the ratios of mono- and diunsaturated CHCs of beetles feeding on these plants by the Mann-Whitney U test.

Previously, we have analyzed the entire CHC profiles of *P. cochleariae* feeding on Chinese cabbage or watercress (Geiselhardt et al. 2012). In the present study, we conducted all statistical analyses separately for straight-chained and methyl-branched CHCs because of their different biosynthesis pathways. First, we examined by a MANOVA whether the quantitative CHC patterns of the beetles depend on sex, host plant species or on interaction of these factors. The same statistical analyses were conducted for CHCs of beetles that fed on the different types of artificial diet.

In addition, canonical discriminant analyses (DA) were performed to determine whether CHC profiles of male and female *P. cochleariae* feeding on different plants or diets could be separated based on the relative composition of CHCs. The DAs were conducted as described in detail by Geiselhardt et al. (2012). To assess the quality of the DAs, the proportion of correctly classified cases was determined by leave-one-out cross validation.

In order to obtain a quantitative measure for the (dis)similarity of male and female CHC profiles, we determined the chemical distances of male CHC profiles to female CHC profiles by the squared Mahalanobis distance. Differences between the chemical distances of CHC profiles of males to the group centroids of females feeding on the same diet or a different diet were analyzed by means of the t-test for paired samples. Prior to the analyses, DAs of CHCs of beetles feeding on the different artificial diets were conducted pairwise by comparing the profile of beetles fed with one type of diet directly with the profile of beetles fed with a single other type.

Differences in mating propensities of beetles fed with different artificial diets were analyzed by χ^2-tests, and differences in copulation duration were analyzed by means of the Kruskal-Wallis H test. Data were corrected for multiple comparisons by using the Bonferroni-Holm-correction. Survival rates of males and females fed with Chinese cabbage or with either type of diet were compared by chi-square tests and Bonferroni-Holm corrected for multiple comparisons. Weight data of males and females fed with Chinese cabbage or with either type of diet were compared by one-way ANOVA (SI Table 4.1).

4.4 Results

Comparison of Fatty Acid Compositions of Host Plant Species. Leaves of Chinese cabbage and watercress contained equal total amounts of fatty acids (Chinese cabbage: 2.3±0.6 mg/g fresh weight; watercress: 2.6±0.8 mg/g fresh weight; Mann-Whitney U test: N = 10, U = 27, P > 0.05). Furthermore, they comprised the same C16- and C18-fatty acids; they did not differ qualitatively with respect to their fatty acid compositions (Table 4.1). Extracts of both plant species were dominated by similar proportions of triunsaturated fatty acids (Chinese cabbage: 69.9±2.2%; watercress: 68.5±0.5%; U = 30, P > 0.05) followed by saturated fatty acids (Chinese cabbage: 14.9±1.3%; watercress: 14.8±0.9%; U = 48, P > 0.05) (Table 4.1). The most abundant compounds were α-linolenic acid (18:3,n-3), palmitic acid (16:0), and roughanic acid (16:3,n-3).

Table 4.1: Mean (± S.D.) relative composition (% total peak area) of the pattern of leaf fatty acids of Chinese cabbage and watercress (N = 10). Relative quantities of fatty acids were analyzed after transesterification into fatty acid methyl esters

Fatty acid[a]	Trivial name	Abbreviation	Chinese cabbage	Watercress
Hexadecanoic acid	Palmitic acid	16:0	13.5±1.3	13.9±0.9
9-Hexadecenoic acid	Palmitoleic acid	16:1,n-7	0.5±0.1	0.4±0.1
7-Hexadecenoic acid	Hypogeic acid	16:1,n-9	2.6±0.3	2.7±0.1
7,10-Hexadecadienoic acid		16:2,n-6	1.0±0.1	0.7±0.2
7,10,13-Hexadecatrienoic acid	Roughanic acid	16:3,n-3	16.1±1.8	15.3±0.9
Octadecanoic acid	Stearic acid	18:0	1.4±0.3	0.9±0.1
11-Octadecenoic acid	Vaccenic acid	18:1,n-7	2.0±0.4	1.4±0.1
9-Octadecenoic acid	Oleic acid	18:1,n-9	1.6±0.4	0.6±0.1
9,12-Octadecadienoic acid	Linoleic acid	18:2,n-6	7.5±0.3	10.9±0.8
9,12,15-Octadecatrienoic acid	α-Linolenic acid	18:3,n-3	53.8±0.9	53.2±1.2
Fatty acid group				
Saturated fatty acids		Cx:0	14.9±1.3	14.8±0.9
Monounsaturated fatty acids		Cx:1	6.7±0.5	5.1±0.3
Diunsaturated fatty acids		Cx:2	8.4±0.2	11.6±1.0
Triunsaturated fatty acids		Cx:3	69.9±2.2	68.5±0.5

[a] all double bonds are in Z-configuration

However, both plant species differed significantly in the relative quantitative composition of their fatty acid patterns (MANOVA, Wilks' λ = 0.02, F = 55.9, P < 0.001). When comparing relative quantities of single fatty acids or single fatty acid groups between plant species (Table 4.1), we detected significant differences between plant species only in their relative proportions of monounsaturated fatty acids (U = 0, P < 0.001) and diunsaturated fatty acids (U = 0, P < 0.001) (Table 4.1). These differences resulted in significantly different ratios of mono- and diunsaturated fatty acids between plant species (U = 0, P < 0.001) (Figure 4.1A).

Impact of Host Plant Species on Insect Straight-Chained CHCs. Feeding on different host plant species did not result in differences of the qualitative composition of insect CHCs. We could confirm previous results and detected the same straight-chained CHCs as those identified previously from cuticular extracts of *Phaedon cochleariae* fed upon Chinese cabbage or watercress (Geiselhardt et al. 2012; SI Table 4.2). The straight-chained CHC fraction of adult beetles feeding on either plant species contained *n*-alkanes and unsaturated hydrocarbons with one (positions 7Z or 9Z), two (6Z,9Z), or three (3Z,6Z,9Z) double bonds (SI Table 4.2).

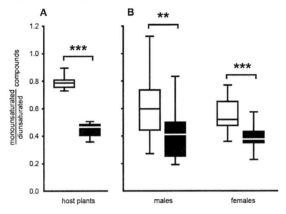

Figure 4.1: (A) Ratios of mono- and diunsaturated fatty acids of Chinese cabbage (white) and watercress (black) (*N* = 10 for each plant species). (B) Ratios of the corresponding straight-chained mono- and diunsaturated hydrocarbons of cuticular extracts of 42-day-old female and male *Phaedon cochleariae* feeding upon the respective host plant species (*N* = 20 for each sex × host plant combination). Boxes indicate 25% and 75% percentiles, bars inside boxes are medians, and whiskers indicate minima and maxima. ** *P* < 0.01 and *** *P* < 0.001 (Mann-Whitney *U* test).

However, the relative composition of the straight-chained CHC fraction (all straight-chained CHCs = 100%) differed significantly according to sex (MANOVA, Wilks' λ = 0.21, *F* = 17.4, *P* < 0.001), host plant species (MANOVA, Wilks' λ = 0.33, *F* = 9.1, *P* < 0.001), and sex × host interaction (MANOVA, Wilks' λ = 0.43, *F* = 6.0, *P* < 0.001).

Furthermore, a DA based on relative compositions of the straight-chained CHC fractions clearly separated males and females kept on the two host plants (Wilks' λ = 0.04, χ^2 = 233.7, *df* = 42, *P* < 0.001; Figure 4.2A), and 86.3% of the cross-validated cases were correctly classified.

Squared Mahalanobis distances showed that the distance between straight-chain CHC patterns of males and females reared on watercress was closer than the one between males on watercress and females on Chinese cabbage; in other words, males reared on watercress had straight-chain CHC profiles more similar to those of "same host" females than to those of "different host" females (*t*-test for paired samples, *t* = -4.9, *P* < 0.001). In contrast, males reared on Chinese cabbage had similar chemical distances to both female types ($t = 0.8$; $P = 0.44$) (Table 4.2; Figure 4.2A).

Interestingly, the ratio of mono- to diunsaturated CHCs of the beetles reflected the ratio of mono- to diunsaturated fatty acids of the host plant species (Figure 4.1B). Beetles of both sexes had a significantly higher monoene to diene ratio when reared on Chinese cabbage compared to the respective sex reared on watercress (males: $N = 20$, $U = 83$, $P < 0.001$; female: $N = 20$, $U = 39$, $P < 0.001$).

Table 4.2: Comparisons of squared Mahalanobis distances (means ± S.E.) from pairwise discriminant analysis based on straight-chained and methyl-branched cuticular hydrocarbons, respectively, of male *Phaedon cochleariae* to the group centroids of females reared either on the same or an alternative (different) plant species or diet[1].

Male	Straight-chained CHCs				Methyl-branched CHCs			
	Female		t^2	P^3	Female		t	P
	Same	Different			Same	Different		
Watercress (W) *versus* Chinese cabbage (C)								
W	43.7±11.9	49.8±13.5	-4.9	**< 0.001**	39.6±22.0	56.7±20.9	-12.6	**< 0.001**
C	29.9±16.0	29.1±17.9	0.8	0.44	46.3±19.3	51.5±20.0	-2.6	**0.02**
Sunflower oil (S) *versus* Linseed oil (L)								
S	20.0±10.9	27.7±13.0	-5.9	**< 0.001**	29.5±12.8	28.8±14.0	0.7	0.47
L	22.9±10.7	24.5±11.9	-1.3	0.21	30.0±12.1	30.8±11.6	-0.9	0.41
Olive oil (O) *versus* Linseed oil (L)								
O	21.4±19.6	30.2±23.2	-7.1	**< 0.001**	34.1±22.5	34.3±23.2	-0.3	0.76
L	18.8±12.0	17.4±12.9	1.2	0.23	22.9±10.9	23.1±12.1	-0.2	0.85
Olive oil (O) *versus* Sunflower oil (S)								
O	24.7±18.6	39.8±21.3	-14.5	**< 0.001**	37.6±24.6	33.5±23.8	4.1	**< 0.001**
S	23.1±10.3	17.9±8.6	4.9	**< 0.001**	23.9±10.6	24.3±11.5	-0.5	0.62

[1] artificial diets differed only in the type of vegetable oil supplemented as fatty acid source. $N = 20$ for all comparisons
[2] *t*-test for paired samples
[3] significant differences ($P < 0.05$) are marked in bold

Impact of Different Artificial Diets on Insect Straight-Chained CHCs. To investigate further the impact of dietary fatty acids on the CHC profiles of beetles, we reared adult beetles on three artificial diets which differed mainly in their fatty acid composition. Feeding on these different types of diet did not result in differences of the qualitative composition of insect CHCs; we detected the same compounds as those detected in beetles feeding on host plants (compare Geiselhardt et al. 2012; SI Table 4.2).

However, the CHC patterns of the beetles differed quantitatively in dependence of the type of diet. After 42 days of feeding on artificial diet, the relative composition of straight-chained CHCs differed significantly according to sex (MANOVA, Wilks' $\lambda = 0.37$, $F = 10.6$, $P < 0.001$) and type of artificial diet (Wilks'

$\lambda = 0.29$, $F = 5.3$, $P < 0.001$); but no sex × diet interaction was observed (Wilks' $\lambda = 0.75$, $F = 0.9$, $P = 0.57$) (all straight-chained CHCs = 100%).

Hence, we performed a DA separately for each sex. The straight-chain CHC profiles of male (Figure 4.3A) and female (Figure 4.3C) beetles were separated according to the type of artificial diet (males: Wilks' $\lambda = 0.38$, $\chi^2 = 65.4$, $P < 0.001$ and females: Wilks' $\lambda = 0.19$, $\chi^2 = 81.4$, $P < 0.001$), and 65.0% and 61.7% of the male and female cases, respectively, were correctly classified by cross validation. The corresponding loading plots indicate that especially unsaturated CHCs contributed to the separation of straight-chained CHC profiles of beetles (males and females) feeding on different types of artificial diets, whereas n-alkanes – diffusely distributed in the plot center – contributes less to the separation of these profiles (Figure 4.3B for males, 4.3D for females).

In order to elucidate whether straight-chained CHC profiles of mating partners feeding on the same artificial diet are more similar to each other than those of mating partners feeding on different artificial diets, we first performed DAs based on straight-chained CHC profiles by running the DAs pairwise for all diet combinations (Figure 4.2) and then calculated the squared Mahalanobis distances between male and female straight-chained CHC profiles (Table 4.2).

When considering the quantitative pattern of straight-chained CHCs, the effect of the dietary fatty acid composition on chemical distances of CHC profiles between males and females reared on same or different diets was not consistent (Table 4.2). CHC profiles of "sunflower oil" males were more similar to those of "same diet" females compared to "linseed oil" females (Figure 4.2C). Males fed on artificial diet supplemented with olive oil had CHC profiles more similar to those of "same diet" females than to "different diet" females (either "linseed oil" or "sunflower oil" females; Figure 4.2E, G). In "linseed oil" males, the chemical distances of straight-chained CHC profiles to those of "same diet" and "different diet" females did not differ in any comparison (Figure 4.2C, E).

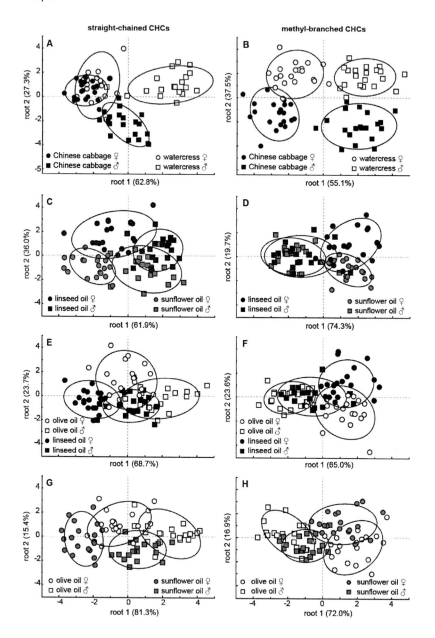

◀ **Figure 4.2:** Results of discriminant analyses based on relative proportions of straight-chained (left row) and methyl-branched (right row) hydrocarbons of cuticle extracts from 42-day-old male (square) and female (circle) *Phaedon cochleariae* beetles reared on different host plant species or artificial diets (N = 20 for each sex × plant/diet combination). Scatterplots of canonical root 1 *vs*. 2. The variance explained by each canonical root is given in parentheses after the respective root. Beetles fed during their adult stage either on (A,B) Chinese cabbage (black) or watercress (white); or on artificial diet supplemented either (C,D) with sunflower oil (grey) or linseed oil (black), (E,F) with olive oil (white) or linseed oil (black), or (G,H) with olive oil (white) or sunflower oil (grey).

Impact of Host Plant Species on Insect Methyl-Branched CHCs. Extracts from beetles feeding on either plant species contained the same methyl-branched CHCs; hence, the methyl-branched beetle CHC patterns did not differ qualitatively in dependence of the plant species (compare with Geiselhardt et al. 2012; SI Table 4.3). However, divergent host plant use clearly affected the quantitative patterns of methyl-branched CHCs of adult *P. cochleariae* (all methyl branched CHCs = 100%). The relative composition of methyl-branched CHCs differed significantly according to sex (MANOVA, Wilks' λ = 0.10, F = 16.3, P < 0.001), host plant species (Wilks' λ = 0.16, F = 8.9, P < 0.001) and sex × host plant interaction (Wilks' λ = 0.41, F = 2.5, P < 0.002).

The DA clearly separated males and females kept on the two host plant species based on their methyl-branched CHC fraction (Wilks' λ = 0.19, χ^2 = 263.8, df = 63, P < 0.001) (Figure 4.2B), and 83.8% of the cross-validated cases were correctly classified. The first canonical root accounted for 55.1% of the total variance of the data and separated beetles according to their sex. The second canonical root, explaining 37.5% of the total variance, clearly separated beetles reared as adults on Chinese cabbage from beetles reared as adults on watercress. Squared Mahalanobis distances showed that CHC profiles of males on both host plant species were more similar to those of "same host" females than to those of "different host" females (*t*-test for paired samples, watercress males: t = -12.6, P < 0.001; Chinese cabbage males: t = -12.2, P = 0.02) (Table 4.2).

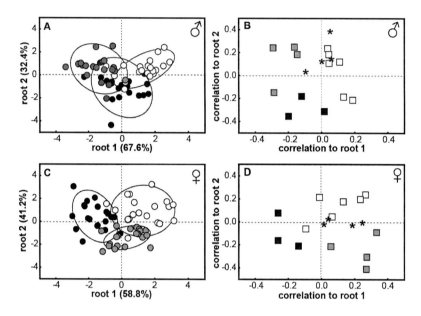

Figure 4.3: Results of a discriminant analysis based on relative proportions of straight-chained hydrocarbons of cuticle extracts from male (A, B) and female (C, D) *Phaedon cochleariae* beetles reared on different artificial diets (N = 20 for each sex × diet combination). A, C: Scatterplots of canonical root 1 *vs.* 2; beetles were fed with diet supplemented with either olive oil (white circle), sunflower oil (grey circle) or linseed oil (black circle). The variance explained by each canonical root is given in parentheses after the respective root. B, D: Corresponding loading plots; mono- (white square), di- (grey square) and triunsaturated (black square) hydrocarbons; *n*-alkanes (asterisk).

Impact of Different Artificial Diets on Insect Methyl-Branched CHCs.
Extracts from beetles feeding on either type of artificial diet contained the same methyl-branched CHCs as those identified from beetles feeding on their host plants (compare Geiselhardt et al. 2012); hence, the methyl-branched beetle CHC patterns did not differ qualitatively in dependence of the type of diet (SI Table 4.3).

However, again quantitative differences were detected between CHC profiles of beetles feeding on the different types of diets. The beetles showed significant differences in the relative composition of their methyl-branched CHCs according to sex (MANOVA, Wilks' λ = 0.31, F = 5.3, P < 0.001), diet (Wilks' λ = 0.37, F = 1.6, P < 0.008), and sex × diet interaction (Wilks' λ = 0.39, F = 1.5, P < 0.02) (all methyl-branched CHC = 100%). Chemical distances between "same diet" and "different diet" couples did not differ when considering the methyl-branched CHC profiles (Table 4.2, Figure 4.2D, F, H), except for "olive oil" males that were more different from "same diet" than to "sunflower oil" females (Table 4.2, Figure 4.2H).

The Influence of Dietary Fatty Acids on Insect Mating Behavior. The composition of the artificial diet had no effect on the mating behavior of *P. cochleariae*. Male mating propensities and the durations of copulation attempts were similar in all male × female combinations regardless of the diet that had been ingested by males and females (Figure 4.4).

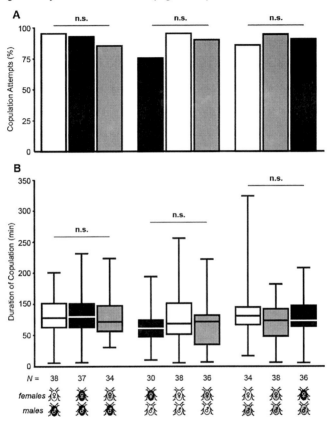

Figure 4.4: Mating propensity of male *Phaedon cochleariae* (A) and duration of copulation attempts (B) with females feeding on different artificial diets supplemented either with olive oil (white), sunflower oil (grey) or linseed oil (black) (*N* = 40 for all combinations). Boxes indicate the 25% and 75% percentiles, the bars inside the boxes are the medians, and whiskers indicate the minima and maxima. n.s. *P* > 0.05, evaluated by means of χ^2-test (A) or Kruskal-Wallis *H* test (B). All data were Bonferroni-Holm-corrected for multiple comparisons.

4.5 Discussion

Our study demonstrates that the composition of ingested fatty acids significantly affects the composition of the CHC profile of adults of the herbivorous mustard leaf beetle *Phaedon cochleariae*. Feeding of beetles on either Chinese cabbage or watercress resulted in different quantitative patterns of both straight-chained and methyl-branched CHCs. Both plant species have similar total fatty acid contents; thus the divergence of the CHC phenotypes depended on differences in the dietary fatty acid composition rather than on different amounts of ingested fatty acids. Beetles feeding on either of the two host plant species that varied in their fatty acid composition (ratio of mono- to diunsaturated fatty acids) showed specific CHC profiles that differed in ratios of the respective straight-chained hydrocarbons. Thus, our data strongly suggest that dietary fatty acids are used as precursors for the biosynthesis of straight-chained CHCs in *P. cochleariae*. Divergence of CHC patterns of beetles was also detected when beetles fed upon artificial diets which differed mainly with respect to their fatty acid compositions. Beetles that fed on these different types of artificial diets showed clear differences in their pattern of straight-chained and methyl-branched CHCs. Hence, ingestion of certain blends of fatty acids may significantly determine the CHC pattern of *P. cochleariae*.

It is striking that the ingested fatty acid composition did not only alter the composition of straight-chained CHCs of adult *P. cochleariae*, but also the composition of methyl-branched CHCs. Based on the current knowledge of the biosynthesis of methyl-branched CHCs, it is unlikely that dietary fatty acids serve as precursors for this class of CHCs as methyl-branches are incorporated early in chain synthesis (Blomquist 2010). Nevertheless, the uptake of fatty acids seems to affect the biosynthesis of methyl-branched CHCs by a yet unknown mechanism. Our findings are corroborated by a study of *Drosophila mojavensis* which showed that the fatty acid composition of a larval diet determined the abundance of a wide range of methyl-branched CHCs in adult flies (Etges et al. 2006).

While alternative host plant use by *P. cochleariae* adults caused a common host-specific signature in the methyl-branched CHC phenotypes of both sexes, the contrasting fatty acid composition in the artificial diets affected only female CHC phenotypes. The CHC phenotypes might not only be influenced by the qualitative fatty acid composition, but also the absolute fatty acid content. However, feeding on different host plant species has a stronger impact on the CHC phenotype than feeding on different artificial diets, albeit the total fatty acid contents of leaves (0.7-3.7 mg/g fresh weight) were somewhat lower than those of artificial diets (approx. 10 mg/g fresh weight). Thus, plant parameters other than fatty acid composition or content are needed to cause a coordinated shift of the methyl-branched CHC phenotypes in males and females. The biosynthesis of methyl-branched CHCs in insects relies on the uptake of some plant-derived essential amino acids, i.e. valine, leucine, isoleucine, and methionine (Blomquist 2010). Valine and leucine are used to initiate the biosynthesis of 2-methylalkanes with even-numbered and odd-numbered carbon backbones, respectively (Blailock et al. 1976). Moreover, valine, isoleucine, and methionine are precursors for methylmalonyl-CoA that gives rise to methyl-branched CHCs (Dwyer et al. 1981; Dillwith et al. 1982; Chase et al. 1990). Although the amino acid pattern in leaves of different plant species are very similar, the leaf protein contents show considerable variation among species (Yeoh et al. 1992). Hence, a different availability of these amino acids in alternative host species might affect the quantities of methyl-branched CHCs and might cause host-specific CHC phenotypes in both sexes. However, differences in the amino acid content could not explain the striking differences in the ratio of 2-MeC28 to 2-MeC30 between beetles that fed on leaves (7.8±3.4; N = 80) and beetles reared on artificial diets (1.4±0.7; N = 120) (SI Table 4.3). Thus, additional factors others than plant-derived precursors may affect the CHC phenotypes by interfering with the regulation of CHC biosynthesis.

When considering how ingested fatty acids might be recruited by the beetles for CHC biosynthesis, the absorbed dietary fatty acids may directly be transported from the midgut tissue to oenocytes which are known to be required for CHC biosynthesis in *Drosophila* (Billeter et al. 2009). However, absorbed fatty acids might also be first stored as triacylglycerides in trophocytes of the fat body (Makki et al. 2014). In our present study, adult *P. cochleariae* fed on different host plant species or on artificial diets with different fatty acid compositions for six weeks. A previous study showed that a two-week feeding period on a novel host was sufficient to induce a significant quantitative change of the beetle´s CHC phenotype (Geiselhardt et al. 2012). Thus, the dietary fatty acids might have been stored in the fat body before they were mobilized and transported to the oenocytes. The storage of fatty acids as triacylglycerides can significantly influence their availability for biosynthesis of CHCs. In Lepidoptera, absorbed fatty acids can be incorporated in different insect lipids; dietary C18:0, C18:2, and C18:3 were mainly recovered in polar insect phospholipids, while neutral insect triacylglycerides contain relative large proportions of C16:0, C16:1, and C18:1 (Grau & Terriere 1971; Turunen 1973). Hence, the dietary fatty acid composition may differ from the composition of potential CHC precursors stored as triacylglycerides in the fat body. However, the finding that the ratios of mono-and diunsaturated fatty acids in watercress and Chinese cabbage matched the ratios of mono- and diunsaturated straight-chained CHCs in male and female *P. cochleariae* suggests that some host plant fatty acids are used as direct precursors by the beetles for their CHC biosynthesis. To corroborate this assumption, stable isotope (^{13}C) labelling of dietary precursors could be used to follow their incorporation into insect pheromones (Dwyer et al. 1981; Dillwith et al. 1982; Blaul & Ruther 2011).

CHCs serve as mate recognition cues in many insect species (Singer 1998; Howard & Blomquist 2005). Diet-induced changes in CHC phenotypes may disrupt mate recognition between males and females with divergent host use and thus, favor ecological speciation by host-specific assortative mating (Fitzpatrick

2012; Geiselhardt et al. 2012). However, although our study has demonstrated that the fatty acid composition of the artificial diets affected both straight-chained and methyl-branched CHCs in *P. cochleariae*, these changes caused no diet-specific assortative mating. Male mating propensities were similar in all male × female combinations irrespective of the type of artificial diet the partners had fed on. In contrast, beetles fed with either Chinese cabbage or watercress clearly showed assortative mating (Geiselhardt et al. 2012).

A previous study has suggested that self-referent phenotype matching (Dawkins 1982; Mateo 2004) is involved in mate recognition in male *P. cochleariae* (Geiselhardt et al. 2012). The current study showed for several couple combinations that the distances between straight-chained male and female CHC profiles were closer (significantly shorter) in "same diet" couples than in "different diet" couples (Table 4.2). In contrast, the distances between methyl-branched male and female CHC profiles hardly differed when comparing "same" with "different" diet couples (except for the "olive oil" vs. "sunflower oil" combination) (Table 4.2). A lack of significant differences in distances between CHC profiles of "same" and "different" diet mating partners indicates that the CHC profiles of "same" diet partners are as different from each other as those of "different" diet partners. Hence, it will be difficult for a male to distinguish between "same" and "different" diet females by self-referent phenotype matching when significant differences in CHC profile distances are lacking; all the females are "different" at the same "distance level of difference".

Even though the straight-chain CHC-profile-distances were significantly different between "same" and "different" diet mating partners (Table 4.2), these differences did obviously not elicit any assortative mating behavior. The straight-chained CHCs of *P. cochleariae* are dominated by alkenes (about 90% of straight chained CHCs) that alone do not trigger a mating response in males; only when adding alkanes to alkenes, can the mixture elicit male mating behavior (Geiselhardt et al. 2009). Even though quantitative changes of the beetle´s cuticular alkenes also change the relative mixture of cuticular alkanes and alkenes, the findings of our study demonstrate that diet-induced changes in the beetle´s straight-chained CHCs (including mainly alkenes) *per se* cannot affect male mating behavior.

The methyl-branched CHC-profile-distances did not significantly differ between most of the combinations of "same" and "different" diet mating partners, whereas significant differences in distances of these profiles were detected for the beetles feeding on Chinese cabbage and watercress (Table 4.2). Hence, our results suggests that females fed with cabbage showed a sufficiently different methyl-branched CHC pattern from females fed with watercress so that males fed with either plant species could distinguish between "same" and "different" female phenotypes. In contrast, females fed with an artificial diet did not show a methyl-branched CHC pattern that was more different from the male pattern than the one of a female fed with another diet; thus, males could not distinguish between these females since these females were both similarly different from the male phenotypes. The presence of differences in distances between these methyl-branched CHC-profiles of males and females feeding on the same and different host plant species and the absence of such differences in beetles feeding on artificial diets indicate that assortative mating behavior does not only require (i) a male´s ability to distinguish between "same" and "different" CHC phenotype females, but also (ii) a female´s ability to synthesize a (methyl-branched) CHC pattern that is on the one hand sufficiently different from a female fed with different food and on the other hand sufficiently similar to the phenotype of a male which ingested the same food; our data suggest that these male and female abilities will allow successful self-referent phenotype matching and thus, assortative mating.

In conclusion, our study clearly demonstrates that ingestion of different fatty acid blends leads to divergent straight-chained and methyl-branched CHC phenotypes of adult *P. cochleariae*. Furthermore, our results showed that this *per se* do not necessarily lead to diet-specific assortative mating since the beetles fed with artificial diets differing in their fatty acid composition did not show diet-specific assortative mating, whereas beetles fed upon different host plant species prefer "same-plant" mates. Thus, future studies need to elucidate which other plant constituents are involved in the alteration of the CHC phenotype in adult *P. cochleariae* that cause host-specific assortative mating (Geiselhardt et al. 2012).

4.6 Supplemental Data

SI Table 4.1: Mean weights (± S.E.) and survival rates of adult *Phaedon cochleariae* feeding for 37 days on Chinese cabbage leaves or artificial diets containing different compositions of triacylglycerides. Stocks originated from a laboratory line feeding on Chinese cabbage until pupation. As control group beetles were offered no diet at all.

Diet	N		Survival rate (%)[*]		Weight (mg)[**]	
	males	females	males	females	males	females
Chinese cabbage leaves	31	21	90.3[a]	76.2[a]	6.6±0.1[a]	9.5±0.4[a]
artificial diet with olive oil[1]	49	66	40.8[b]	37.9[b]	6.3±0.2[a]	8.0±0.2[b]
artificial diet with sunflower oil[2]	62	54	20.9[b]	40.7[b]	6.0±0.2[a]	8.2±0.3[b]
artificial diet with linseed oil[3]	61	62	21.3[b]	43.6[b]	6.1±0.2[a]	8.0±0.3[b]
no diet	21	21	0[c]	0[c]	-	-

[*] different letters in each column indicate significant differences by chi-square test ($P \leq 0.05$). Data were Bonferroni-Holm-corrected for multiple comparisons.
[**] different letters in each column indicate significant differences by one-way ANOVA ($P \leq 0.05$).
[1]containing mainly monounsaturated fatty acids
[2]containing mainly diunsaturated fatty acids
[3]containing mainly triunsaturated fatty acid

SI Table 4.2: Mean (± S.D.) relative composition (% total peak area) of straight-chained cuticular hydrocarbons (=100%) of male and female *Phaedon cochleariae* feeding on different artificial diets containing different amounts of triglycerides and Chinese cabbage or watercress (*N* = 20 per sex and diet). Different stocks originated from a laboratory line feeding until pupation on Chinese cabbage. Crude extracts (dichloromethane) were analyzed by GC/MS (DB-5).

Compound	M+-Ion	Chinese cabbage		Watercress		Linseed oil		Olive oil		Sunflower oil	
		Males	Females	Males	Females	Males	Females	Males	Females	Males	Females
9-C23ene	322	28.1±6.8	21.8±3.7	20.7±6.2	15.9±2.6	13.7±4.9	5.2±2.8	15.4±6.1	9.3±6.5	10.1±3.9	3.5±1.7
6,9-C23diene	320	52.7±8.1	44.0±5.8	58.8±7.8	47.7±5.2	43.8±13.6	24.0±13.7	42.5±9.7	33.1±13.6	49.2±14.1	29.8±10.9
C23triene	318	0.1±0.3	1.0±0.8	0.3±0.5	0.3±0.3	0.3±0.3	0.3±0.3	0.1±0.2	0.1±0.2	0.1±0.2	0.1±0.1
n-C25	352	2.5±1.2	3.7±1.5	1.8±1.2	2.4±1.2	1.8±1.8	2.6±2.9	2.6±2.6	2.5±1.8	1.6±1.5	2.1±1.3
9-C25ene	350	0.3±0.5	0.6±0.6	0.5±0.5	0.8±0.7	1.0±0.8	1.5±1.2	0.7±0.5	0.8±0.6	0.5±0.5	0.8±1.0
7-C25ene	350	0.9±0.7	3.2±1.5	0.2±0.6	2.9±1.7	3.5±1.4	5.9±5.5	2.7±1.2	2.8±1.5	2.2±1.1	2.9±1.7
6,9-C25diene	348	1.0±0.9	4.8±2.8	1.4±1.0	5.9±2.5	4.6±2.9	5.1±2.7	2.4±1.4	4.7±1.7	3.2±1.6	6.4±3.4
3,6,9-C25triene	346	0.2±0.4	2.8±1.6	0.4±0.7	1.6±1.5	2.4±1.6	4.0±2.0	1.3±0.9	2.0±1.3	1.5±1.1	3.1±2.3
n-C27	380	2.0±0.7	2.1±0.7	2.7±1.0	2.2±1.0	1.5±1.6	2.8±2.7	1.5±1.7	1.8±1.1	1.5±1.7	1.7±1.3
9-C27ene	378	2.1±1.7	2.5±1.8	2.1±1.8	3.4±1.6	4.0±1.6	6.2±2.7	4.9±2.4	4.6±3.1	3.6±1.8	4.5±1.6
7-C27ene	378	1.8±1.7	0.8±0.7	1.2±1.1	0.7±0.7	1.6±0.9	1.9±1.3	1.6±0.7	1.6±0.9	1.3±0.6	1.2±0.7
6,9-C27diene	376	3.7±2.4	5.4±2.4	4.5±2.7	8.2±2.0	9.2±3.3	16.7±5.4	9.1±6.7	17.6±9.5	10.7±5.0	25.9±8.5
3,6,9-C27triene	374	0.9±1.3	5.1±2.0	0.2±0.4	3.7±1.5	5.6±3.8	12.7±7.1	2.5±1.7	5.2±2.7	3.2±2.5	7.4±3.5
n-C28	394	1.8±0.9	1.0±0.6	3.6±1.8	2.6±1.0	2.6±3.9	4.1±4.9	6.1±5.6	6.3±7.2	4.1±4.8	3.5±2.9
n-C29	408	2.0±0.7	1.2±0.4	1.8±1.5	1.8±1.1	2.9±3.7	4.9±3.7	3.7±4.6	3.8±2.9	3.5±4.9	3.8±3.1
C29ene	318	-	-	-	-	0.6±0.8	0.4±0.7	1.2±1.2	2.1±4.7	1.0±1.4	0.3±0.5
6,9-C29diene	352	-	-	-	-	1.2±1.3	1.9±2.1	1.8±1.5	1.7±1.5	2.9±3.0	3.2±2.1
Cx:0		33.1±7.4	28.9±3.8	24.6±7.0	23.6±3.6	8.7±9.5	14.4±12.6	13.9±12.4	14.4±11.2	10.6±12.6	11.0±7.5
Cx:1		57.4±7.8	54.3±4.5	64.7±8.1	61.8±3.5	24.4±6.0	21.0±9.1	26.5±6.0	21.2±8.8	18.7±4.4	13.1±3.3
Cx:2		1.2±1.3	8.8±2.7	0.9±1.2	5.6±2.2	58.8±12.1	47.6±16.1	55.7±11.2	57.2±10.6	65.9±13.2	65.3±11.5
Cx:3		8.3±2.4	8.0±2.8	9.9±4.3	9.0±3.9	8.2±4.8	16.9±8.3	3.9±2.4	7.3±3.7	4.8±3.5	10.6±5.2

SI Table 4.3: Mean (± S.D.) relative composition (% total peak area) of methyl-branched cuticular hydrocarbons of male and female Phaedon cochleariae feeding on different artificial diets containing different amounts of triglycerides and Chinese cabbage or watercress (N = 20 per sex and diet). Different stocks originated from a laboratory line feeding until pupation on Chinese cabbage. Crude extracts (dichloromethane) were analyzed by GC/MS (DB-5).

Compound	Chinese cabbage Males	Chinese cabbage Females	Watercress Males	Watercress Females	Linseed oil Males	Linseed oil Females	Olive oil Males	Olive oil Females	Sunflower oil Males	Sunflower oil Females
11/13-MeC23	0.4±0.1	0.4±0.2	0.1±0.1	0.2±0.2	-	-	-	-	-	-
2-MeC24	6.0±1.7	14.5±3.1	4.3±1.1	10.8±3.4	4.6±2.5	3.1±1.7	3.8±3.4	3.8±2.0	4.1±2.1	2.7±1.5
13-MeC25-7-ene / 15-MeC25-7-ene	0.6±0.2	1.1±0.5	0.2±0.2	0.7±0.3	0.5±0.2	0.4±0.3	0.5±0.4	0.4±0.4	0.5±0.3	0.3±0.2
2-MeC25	0.4±0.1	0.6±0.1	0.3±0.1	0.5±0.2	-	-	-	-	-	-
7,13-diMeC25	1.5±0.3	1.4±0.5	1.4±0.3	1.7±0.5	-	-	-	-	-	-
5,13-diMeC25	0.6±0.2	0.5±0.2	0.4±0.2	0.5±0.2	-	-	-	-	-	-
2-MeC26	14.4±2.8	17.2±2.1	12.9±2.5	13.3±2.3	3.6±1.7	2.6±1.4	3.3±1.8	3.4±2.0	4.4±2.7	1.8±0.8
13-MeC27-2-ene / 15-MeC27-4/9-ene	2.6±0.9	4.1±1.9	1.7±0.7	3.5±1.4	2.2±0.6	1.7±1.0	2.2±0.9	1.5±0.8	1.9±1.1	1.4±0.5
13-MeC27	3.5±0.8	4.6±2.2	4.2±1.5	5.7±2.0	1.9±0.9	1.1±0.5	1.3±0.9	1.4±0.9	1.6±0.5	1.3±0.8
2-MeC27 / 9,13-/9,15-diMeC27	1.5±0.2	1.3±0.2	1.4±0.2	1.3±0.2	0.7±0.3	0.3±0.2	0.7±0.4	0.4±0.2	0.8±0.3	0.3±0.2
5,13-/5,15-diMeC27	2.9±0.8	3.1±1.0	2.4±0.5	3.0±0.9	1.1±0.7	0.5±0.2	1.3±0.9	1.0±0.5	1.3±0.8	0.6±0.3
2-MeC28	40.3±6.2	28.1±5.0	44.2±6.7	32.2±6.6	28.6±3.7	27.7±5.8	28.8±7.7	26.5±8.0	29.4±7.7	23.6±6.8
13/15-MeC29	2.6±0.9	2.2±0.7	1.8±0.5	1.8±0.4	0.7±0.6	0.3±0.3	0.6±0.5	0.5±0.2	0.7±0.4	0.3±0.2
2-MeC29	-	-	-	-	0.8±0.2	0.6±0.3	0.7±0.2	0.7±0.2	0.7±0.1	0.7±0.3
2-MeC30	4.6±1.9	3.4±1.2	7.3±3.6	6.9±3.1	20.5±6.3	24.5±6.3	25.1±11.1	21.2±6.4	19.9±5.2	26.3±6.1
12/13-/14-MeC31	1.3±0.5	0.7±0.4	0.8±0.4	0.6±0.3	0.8±0.6	0.4±0.2	0.6±0.5	0.5±0.3	0.6±0.2	0.4±0.3
13-MeC33	0.6±0.3	0.5±0.2	0.5±0.2	0.3±0.3	0.6±0.5	0.3±0.2	0.5±0.3	0.3±0.2	0.4±0.2	0.3±0.2
5,13-diMeC33	1.0±0.3	0.8±0.3	1.0±0.2	0.7±0.4	1.2±0.7	0.9±0.3	1.2±0.6	1.1±0.6	1.3±0.6	0.9±0.4
21-MeC35-10/12-ene / 23-MeC35-12/14-ene	2.7±0.7	1.5±0.5	2.6±0.7	1.3±0.6	-	-	-	-	-	-
13-/15-MeC35	0.5±0.3	0.4±0.3	0.4±0.3	0.4±0.3	0.9±0.4	0.5±0.2	0.7±0.3	0.6±0.3	0.6±0.3	0.6±0.3
7,13-/7,15-diMeC35	-	-	-	-	0.2±0.2	0.1±0.1	0.2±0.2	0.2±0.2	0.2±0.2	0.2±0.2
5,13-/5,15-diMeC35	0.7±0.3	0.5±0.4	0.7±0.3	0.5±0.4	1.6±0.5	1.5±0.4	1.6±0.7	1.6±0.5	2.0±1.2	1.8±0.6
23-MeC36-12-ene / 24-MeC36-13-ene / 25-MeC36-14-ene	0.7±0.4	0.4±0.4	0.5±0.4	0.4±0.3	0.7±0.4	0.6±0.3	0.8±0.5	0.6±0.4	1.0±0.7	0.7±0.3
23-MeC37-12/14-ene / 25-MeC37-14/16-ene	2.3±0.6	1.6±0.6	2.4±0.5	1.7±0.8	1.9±0.6	2.7±1.7	2.0±1.0	1.9±0.7	1.8±0.4	2.6±1.4
25-MeC38-14-ene / 26-MeC38-15-ene / 27-MeC38-16-ene	-	-	-	-	0.7±0.4	0.7±0.3	0.6±0.3	0.7±0.4	0.7±0.4	0.7±0.4
25-MeC39-14/16-ene / 27-MeC39-16/18-ene	2.3±0.8	2.7±1.0	2.3±0.9	3.2±1.7	7.6±2.8	12.0±3.9	6.3±2.9	11.8±5.0	7.3±2.5	13.1±4.6
27-MeC40-16-ene / 28-MeC40-17-ene / 29-MeC40-18-ene	-	-	-	-	0.3±0.3	0.4±0.2	0.4±0.2	0.3±0.2	0.3±0.3	0.5±0.3
27-MeC41-16/18-ene / 29-MeC41-18/20-ene	1.6±0.7	1.5±0.7	1.6±0.8	1.6±0.9	5.1±1.4	5.8±1.6	3.8±1.8	5.9±2.3	4.3±1.9	6.4±1.9
29-MeC42-18-ene / 30-MeC42-19-ene /	-	-	-	-	0.3±0.3	0.5±0.2	0.4±0.3	0.5±0.4	0.4±0.4	0.5±0.3
29-MeC43-18/20-ene / 31-MeC43-20/22-ene	1.8±1.1	1.7±0.9	1.9±0.9	1.6±1.2	5.2±1.6	4.6±2.4	5.2±1.5	5.2±2.1	5.4±1.6	4.6±1.9
31-MeC45-20/22-ene / 33-MeC45-22/24-ene	-	-	-	-	1.1±0.6	0.8±0.4	1.2±0.8	1.1±0.8	1.1±0.6	1.0±0.5

4.7 References

Billeter J-C., Atallah J., Krupp J.J., Millar J.G. & Levine J.D. (2009). Specialized cells tag sexual and species identity in *Drosophila melanogaster*. *Nature*, 461, 987-991.

Blailock T.T., Blomquist G.J. & Jackson L.L. (1976). Biosynthesis of 2-methylalkanes in the crickets *Nemobius fasciatus* and *Gryllus pennsylvanicus*. *Biochemical and Biophysical Research Communications*, 68, 841-849.

Blaul B. & Ruther J. (2011). How parasitoid females produce sexy sons: a causal link between oviposition preference, dietary lipids, and mate choice in *Nasonia*. *Proceedings of the Royal Society B: Biological Sciences*, 278, 3286-3293.

Blomquist G.J. (2010). Biosynthesis of cuticular hydrocarbons, pp. 35-52, in G.J Blomquist and A.G. Bagnères (eds.). Insect Hydrocarbons: Biology, Biochemistry, and Chemical Ecology. *Cambridge University Press, Cambridge, UK*.

Boughman J.W. (2001). Divergent sexual selection enhances reproductive isolation in sticklebacks. *Nature*, 411, 944-948.

Boughman J.W. (2002). How sensory drive can promote speciation. *Trends in Ecology & Evolution*, 17, 571-577.

Brazner J.C. & Etges W.J. (1993). Pre-mating isolation is determined by larval rearing substrates in cactophilic *Drosophila mojavensis*. II. Effects of larval substrates on time to copulation, mate choice and mating propensity. *Evolutionary Ecology*, 7, 605-624.

Browse J., McCourt P. & Somerville C. (1986). Fatty-acid composition of leaf lipids determined after combined digestion and fatty-acid methyl-ester formation from fresh tissue. *Analytical Biochemistry*, 152, 141-145.

Chase J., Jurenka R.A., Schal C., Halarnkar P.P. & Blomquist G.J. (1990). Biosynthesis of methyl branched hydrocarbons in the German cockroach *Blattella germanica* (L.) (Orthoptera, Blattellidae). *Insect Biochemistry*, 20, 149-156.

Coyne J.A. & Orr H.A. (2004). Speciation. *Sinauer, Sunderland.*

Dawkins R. (1982). The Extended Phenotype. *W.H. Freeman, San Francisco.*

Dillwith J.W., Nelson J.H., Pomonis J.G., Nelson D.R. & Blomquist G.J. (1982). A 13C-NMR study of methyl-branched hydrocarbon synthesis in the housefly. *Journal of Biological Chemistry*, 257, 11305-11314.

Dwyer L.A., Blomquist G.J., Nelson J.H. & Pomonis J.G. (1981). A 13C-NMR study of the biosynthesis of 3-methylpentacosane in the American cockroach. *Biochimica et Biophysica Acta*, 663, 536-544.

Egan S.P. & Funk D.J. (2009). Ecologically dependent postmating isolation between sympatric host forms of *Neochlamisus bebbianae* leaf beetles.

Proceedings of the National Academy of Sciences of the United States of America, 106, 19426-19431.

Etges W.J. (1992). Premating isolation is determined by larval rearing substrates in cactophilic *Drosophila mojavensis*. *Evolution*, 46, 1945-1950.

Etges W.J., Veenstra C.L. & Jackson L.L. (2006). Premating isolation is determined by larval rearing substrates in cactophilic *Drosophila mojavensis*. VII. Effects of larval dietary fatty acids on adult epicuticular hydrocarbons. *Journal of Chemical Ecology*, 32, 2629-2646.

Fitzpatrick B.M. (2012). Underappreciated consequences of phenotypic plasticity for ecological speciation. *International Journal of Ecology*, 2012, Article ID 256017, pp. 12.

Francis G.W. & Veland K. (1981). Alkylthiolation for the determination of double-bond positions in linear alkenes. *Journal of Chromatography*, 219, 379-384.

Geiselhardt S., Otte T. & Hilker M. (2009). The role of cuticular hydrocarbons in male mating behavior of the mustard leaf beetle, *Phaedon cochleariae* (F.). *Journal of Chemical Ecology*, 35, 1162-1171.

Geiselhardt S., Otte T. & Hilker M. (2012). Looking for a similar partner: host plants shape mating preferences of herbivorous insects by altering their contact pheromones. *Ecology Letters*, 15, 971-977.

Grau P.A. & Terriere L.C. (1971). Fatty acid profile of the cabbage looper, *Trichoplusia ni*, and the effect of diet and rearing conditions. *Journal of Insect Physiology*, 17, 1637-1649.

Howard R.W. & Blomquist G.J. (2005). Ecological, behavioral, and biochemical aspects of insect hydrocarbons. *Annual Review of Entomology*, 50, 371-393.

Kühbandner S., Hacker K., Niedermayer S., Steidle J.L.M. & Ruther J. (2012). Composition of cuticular lipids in the pteromalid wasp *Lariophagus distinguendus* is host dependent. *Bulletin of Entomological Research*, 102, 610-617.

Maan M.E. & Seehausen O. (2011). Ecology, sexual selection and speciation. *Ecology Letters*, 14, 591-602.

Makki R., Cinnamon E. & Gould A.P. (2014). The development and functions of oenocytes. *Annual Review of Entomology*, 59, 405-425.

Mateo J.M. (2004). Recognition systems and biological organization: the perception component of social recognition. *Annales Zoologici Fennici*, 41, 729-745.

Matsubayashi K.W., Ohshima I. & Nosil P. (2010). Ecological speciation in phytophagous insects. *Entomologia Experimentalis et Applicata*, 134, 1-27.

Millar J.G. (2000). Polyene hydrocarbons and epoxides: A second major class of lepidopteran sex attractant pheromones. *Annual Review of Entomology*, 45, 575-604.

Millar J.G. (2010) Chemical synthesis of insect cuticular hydrocarbons, pp. 163-186, in G.J Blomquist and A.G. Bagnères (eds.). Insect Hydrocarbons: Biology, Biochemistry, and Chemical Ecology. *Cambridge University Press, Cambridge, UK.*

Nosil P. (2012). Ecological speciation. *Oxford University Press, Oxford.*

Nosil P., Vines T.H. & Funk D.J. (2005). Perspective: reproductive isolation caused by natural selection against immigrants from divergent habitats. *Evolution*, 59, 705-719.

Pennanec'h M., Bricard L., Kunesch G. & Jallon J.M. (1997). Incorporation of fatty acids into cuticular hydrocarbons of male and female *Drosophila melanogaster. Journal of Insect Physiology*, 43, 1111-1116.

Rundle H.D. & Nosil P. (2005). Ecological speciation. *Ecology Letters*, 8, 336-352.

Safran R.J., Scordato E.S., Symes L.B., Rodríguez R.L. & Mendelson T.C. (2013). Contributions of natural and sexual selection to the evolution of premating reproductive isolation: a research agenda. *Trends in Ecology & Evolution*, 28, 643-650.

Schluter D. (2001). Ecology and the origin of species. *Trends in Ecology & Evolution*, 16, 372-380.

Schluter D. (2009). Evidence for ecological speciation and its alternative. *Science*, 323, 737-741.

Scribe P., Guezennec J., Dagaut J., Pepe C. & Saliot A. (1988). Identification of the position and the stereochemistry of the double bond in monounsaturated fatty acid methyl esters by gas chromatography/mass spectrometry of dimethyl disulfide derivatives. *Analytical Chemistry*, 60, 928-931.

Sharon G., Segal D., Ringo J.M., Hefez A., Zilber-Rosenberg I. & Rosenberg E. (2010). Commensal bacteria play a role in mating preference of *Drosophila melanogaster. Proceedings of the National Academy of Sciences of the United States of America*, 107, 20051-20056.

Singer T. (1998). Roles of hydrocarbons in the recognition systems of insects. *American Zoologist*, 38, 394-405.

Smadja C. & Butlin R.K. (2009). On the scent of speciation: the chemosensory system and its role in premating isolation. *Heredity*, 102, 77-97.

Stanley-Samuelson D.W., Jurenka R.A., Cripps C., Blomquist G.J. & de Renobales M. (1988). Fatty acids in insects: Composition, metabolism and biological significance. *Archives of Insect Biochemistry and Physiology*, 9, 1-33.

The Marie Curie SPECIATION Network (2012). What do we need to know about speciation? *Trends in Ecology & Evolution*, 27, 27-39.

Turunen S. (1973). Utilization of fatty acids by *Pieris brassicae* reared on artificial and natural diets. *Journal of Insect Physiology*, 19, 1999-2009.

Via S. (2009). Natural selection in action during speciation. *Proceedings of the National Academy of Sciences of the United States of America*, 106, 9939-9946.

Wilkins M.R., Seddon N. & Safran R.J. (2013). Evolutionary divergence in acoustic signals: causes and consequences. *Trends in Ecology & Evolution*, 28, 156-166.

Yeoh H-H., Wee Y-C. & Watson L. (1992). Leaf protein contents and amino acid patterns of dicotyledonous plants. *Biochemical Systematics and Ecology*, 20, 657-663.

5

Phenotypic Plasticity of Cuticular Hydrocarbon Profiles in Insects

5.1 Abstract

Phenotypic plasticity enables organisms to change specific traits in response to environmental conditions. If phenotypic divergence is maintained, it may promote genetic divergence and thus speciation. A deeper understanding of the mechanisms and consequences of phenotypic plasticity will help to elucidate speciation processes. In this review, we focus on the cuticular hydrocarbon (CHC) phenotypes of insects, which are highly plastic in response to both abiotic and biotic factors. The CHCs serve as protectants against detrimental environmental conditions, but also as cues in insect chemical communication. In a first step, we provide an overview of the factors known to shape an insect's CHC phenotype. Further, we address the dynamics of insect CHC phenotypes in response to gradual or saltatory environmental changes. Finally, we discuss the evolutionary consequences of phenotypic plasticity of insect CHC profiles.

5.2 Introduction

Organisms are exposed to a variety of different environmental conditions that may change in space, time, intensity and quality, thus challenging organisms to be prepared for adaptation to a wide range of conditions. The response of an individual to novel environmental conditions can take many forms, ranging from changes in physiology, alterations of morphology to shifts in behavioral responses (Schlichting & Pigliucci 1998; Wund 2012). This variation in phenotypic expression is defined as "phenotypic plasticity" – the ability of a single genotype to produce different phenotypes in response to different abiotic and biotic environmental conditions (Agrawal 2001; Pfennig et al. 2010; Moczek et al. 2011).

Phenotypic plasticity of a certain genotype plays a role in many evolutionary processes like selection within and between species (Salamin et al. 2010), formation of host races (Drès & Mallet 2002) or the build-up of reproductive isolation barriers between and within populations which can reduce the gene flow (Coyne & Orr 2004). Thus, phenotypic plasticity may promote speciation processes (Agrawal 2001; Miner et al. 2005; Pfennig et al. 2010), and facilitate or even speed up the process of (adaptive) evolution (West-Eberhard 2003).

In order to understand how an organism copes with a variety of ancestral and novel environmental conditions, we need to investigate how an organism can respond to changing environmental factors and what are the costs and/or benefits of these phenotypic changes (Pigliucci 2005; Moczek 2010; Snell-Rood 2012).

In this review, we focus on the phenotypic plasticity of cuticular hydrocarbons (CHCs) in insects (Thomas & Simmons 2011) and address especially evolutionary ecology aspects of insect CHC plasticity. Knowledge on insect CHC biochemistry and biosynthesis has excellently been reviewed by Howard & Blomquist (2005) and Blomquist (2010). Special emphasis will be paid here to CHCs of solitary insects rather than of social insects.

In general, CHCs of insects are known to serve three main functions. First, they function as a barrier to water loss and avoid desiccation, second, they protect insects from infections and severe sun radiation, and finally CHCs play an important role in intra- and interspecific communication within and between insect populations (Howard & Blomquist 2005; Blomquist & Bagnères 2010). Known behavioral functions of CHCs comprise the mediation of recognition, aggregation, dispersal, alarm and sexual behavior in insects (Howard 1993; Tillman et al. 1999; Blomquist & Bagnères 2010). While social insects also use CHCs for recognition and interaction with nestmates and as fertility and dominance signals (Singer 1998; Liebig 2010), solitary insects mainly use CHCs for the discrimination of conspecifics, recognition of enemies or hosts, location of mating partners, and elicitation of courtship behavior (Howard & Blomquist 2005; Geiselhardt et al. 2009; Ruther et al. 2011). The behavior-eliciting effect of a CHC profile may be limited to distinct (groups of) CHCs (Bonavita-Cougourdan et al. 1987; Dani et al. 2005; Lucas et al. 2005; Brandt et al. 2009).

CHC profiles (or phenotypes) are typically complex mixtures of n-alkanes, alkenes and methyl-branched alkanes (Martin & Drijfhout 2009). The number, chain length and chemical families of hydrocarbons vary among species, thus providing often unique profiles. Within species, different populations usually have the same (or a similar) blend of CHCs but differ in their relative quantitative chemical composition (Blomquist & Bagnères 2010; van Wilgenburg et al. 2011). There is enormous diversity in the chemical composition of these CHCs, but the underlying mechanism by which this diversity has evolved and the ecological processes that drive this evolution remain poorly understood.

Insect CHC phenotypes are shaped by genetic and environmental factors (Ferveur 2005; Howard & Blomquist 2005; Martin & Drijfhout 2009). Several studies suggest that CHC profiles largely depend on the genetic background (Beye et al. 1998; Van Zweden et al. 2009). Moreover, also endogenous factors, such as endocrine or enzymatic factors, the regulation of transport pathways affect the CHC phenotype. However, the genetically fixed and physiologically determined CHC phenotype is also strongly affected by numerous abiotic and biotic environmental factors.

Here we first will provide an overview of the abiotic and biotic factors, which significantly influence the expression of insect CHC phenotypes. Second, we consider the *"dynamics of phenotypic change"* and address how rapidly and/or temporally a phenotypic change may take place. We further analyze two *"modes of phenotypic change"* and differentiate between gradual and saltatory changes. Finally, we critically discuss *"consequences of phenotypic plasticity"* and address costs and benefits of the plasticity of insect CHC profiles as well as consequences of plasticity of CHC phenotypes in insect chemical communication systems, speciation processes and evolutionary biology.

5.3 Abiotic and biotic factors driving CHC plasticity

Temperature. Climate changes may play a major role in the evolution of different CHC phenotypes. For example, CHCs prevent desiccation, and thus, may be important for species living in warm and dry environments (Rouault et al. 2004). The melting temperature of CHCs increases with increasing chain length, but decreases with the presence of double bonds and methyl-branches. The permeability of CHCs to water can decrease with increasing chain length (Gibbs 1998; Gibbs 2002; Gibbs & Rajpurohit 2010). Furthermore, *n*-alkanes provide qualitatively better waterproofing than *n*-alkenes and methyl-branched alkanes, because they pack well together in a crystalline state (Gibbs & Rajpurohit 2010). Therefore, proportional increase in chain length of *n*-alkanes and decrease in *n*-alkenes and methyl-branched alkanes within a CHC phenotype might serve to reduce water loss by insects exposed to warm and dry conditions (van Wilgenburg et al. 2011). In addition, seasonal variation in the CHC profile might occur due to extrinsic factors (e.g. overwintering). The seasonal increase in CHCs of the tephritid fly *Eurosta solidaginis* may explain the extreme resistance of overwintering larvae to desiccation (Nelson & Lee 2004).

The effects (shift of chain length) of temperatures (especially high temperatures) on insect CHC phenotypes have been demonstrated in several insect orders, for example, Diptera (Toolson 1982; Gibbs et al. 1998; Noorman & den Otter 2002; Rouault et al. 2004), Isoptera (Woodrow et al. 2000), Hymenoptera (Wagner et al. 2001), Coleoptera (Hadley 1977; Geiselhardt et al. 2006), and Orthoptera (Gibbs & Mosseau 1994).

Relative humidity. Alongside with temperature also relative humidity has a great impact on the CHC profile of an insect. Loss of water through the insect cuticle can be expected to be lower at higher relative humidity. Therefore, smaller amounts of CHCs may be needed to protect an insect from water loss. For example, *Musca domestica* flies reared for 3 days at high relative humidity (90%) produce lower amounts of saturated and unsaturated CHCs than those reared at 20% and 50% relative humidity (Noormann & den Otter 2002). Studies of Hymenoptera species show that low relative humidity (dry conditions) results in higher amounts of cuticular *n*-alkanes (Wagner et al. 2001). Hence, both relative humidity and temperature may strongly affect the production of CHCs of insects (Woodrow et al. 2000).

Age. The quantity and quality of a CHC profile may change with the age of an insect. For example, the CHC profiles of puparia and adults of the fly *Lucilia cuprina* show significant qualitative and quantitative differences (Goodrich 1970). The age-dependent change of the CHC profile of an insect is obviously a common phenomenon in insects. Changes in the CHC pattern of various insect species are found throughout the life cycle of other flies, e.g. *Sarcophaga bullata* (Armold & Regnier 1975), *Calliphora vomitoria* (Trabalon et al. 1992; Roux et al. 2006), *Chrysomya rufifacies* (Zhu et al. 2006), *M. domestica* (Gibbs et al. 1995; Mpuru et al. 2001), of mosquitoes, e.g. *Aedes aegypti* (Hugo et al. 2006), of Orthoptera, e.g. *Chorthippus parallelus* (Tregenza et al. 2000), of Lepidoptera, e.g. *Diatraea saccharalis* (Girotti et al. 2012), or of Coleoptera, e.g. *Curculio caryae* (Espelie & Payne 1991). All these studies suggest that CHCs are a useful indicator of age.

Furthermore, changes in the CHC profiles during ageing can influence the behavior, e.g. by decreasing the sexual attractiveness in *Drosophila melanogaster* (Kuo et al. 2012) and *Lariophagus distinguendus* (Kühbandner et al. 2012). Hence, in those species in which CHCs serve as mating signals, the effects of aging on CHC profiles may reflect the reproductive status and might be perceived by potential mates as reduced sexual attractiveness (Kuo et al. 2012).

Interspecific interactions. Besides the above described factors, interactions between insects and other organisms can also influence their CHC phenotype. Below, we will specifically address interactions between (social) parasites and (social) insects and between pathogenic or beneficial microorganisms and insects.

Social insect parasites invade the colony of a social insect and spend a part of their life in this host colony. In order to prevent attack by host colony members, the parasites can change their CHC profile after invasion to match the host CHC profile (Lenoir et al. 2001). This ability of chemical mimicry has been demonstrated in numerous social parasites. It may be due to either transfer of CHCs from one species to the other by social contact or to *de novo* synthesis of the CHC pattern by the invading species (Vander Meer & Wojcik 1982; Dettner &

Liepert 1994; Soroker et al. 1995; Akino et al. 1999; Johnson et al. 2001; Lenoir et al. 2001; Elmes et al. 2002 Pekár & Jiros 2011).

In addition, not only may the CHC profile of a social parasite change when invading the colony of a social insect, but also the CHC pattern of the host insect may change when its body is directly attacked by an ecto- or endoparasite. For example, CHC profiles of non-parasitized *Apis mellifera* and those of honeybees parasitized by the ectoparasitic mite *Varroa jacobsoni* are qualitatively similar, but show significant differences in the relative proportions of hydrocarbons (Salvy et al. 2000). Such differences in the CHC phenotype between non-parasitized and parasitized honeybees may be regarded as a consequence of stress leading to quantitative and/or qualitative modification of the CHC profile. Not only social parasites and ectoparasitoids, but also endoparasitoids are known to modify the CHC profile of their hosts, as was shown for e.g. parasitized ants (e.g. Trabalon et al. 2000).

Furthermore, infection by entomopathogens can change an insect's CHC profile (Pedrini et al. 2013). For example, treatment of male mealworm beetles, *Tenebrio molitor*, with lipopolysaccharides (LPS), which mimic bacterial infection, increases their immune activity and also leads to a higher proportion of long-chain CHCs. Males with increased immune activity are more attractive to females that might use the male CHC profiles to evaluate the immune state of their partner (Nielsen & Holman 2012). In contrast, LPS treatment of the queen of the ant *Lasius niger* reduces the quantities of a specific CHC that also serves as queen pheromone; this effect might be due to a trade-off between immunity and pheromone production (Holman et al. 2010). Furthermore, the total amounts of CHCs in the moth *Ostrinia nubilalis* and the cockchafer *Melolontha melolontha* are significantly reduced during infection with a pathogen (Lecuona et al. 1991).

In *D. melanogaster*, not pathogenic, but even symbiotic bacteria affect the expression of the CHC phenotype (Sharon et al. 2010).

Physiological state. The physiological state of an insect may also affect the expression of its CHC phenotype and may thus affect breeding and sexual behavior. For example, the CHC profiles of both males and females of the burying beetle *Nicrophorus vespilloides* are highly complex and change their composition with the breeding status. Thus, CHCs can inform *Nicrophorus* beetles about the parental state and enable them to discriminate between their breeding partner and a conspecific intruder, but also mediate recognition of breeding partners (Steiger et al. 2007; Scott et al. 2008; Steiger et al. 2008).

In other insects, like the staphylinid beetle *Aleochara curtula*, immature, starved, or multiply mated males mimic their females chemically to avoid intrasexual aggression (Peschke 1985; 1987). In addition to this example of the impact of CHCs on the sexual behavior of a non-social insect, numerous studies show that the CHC profiles of social insects change in dependence of the status of ovary development (fertility) or of mating (Peeters et al. 1999; Liebig et al. 2000; Cuvillier-Hot et al. 2001; de Biseau et al. 2004; Lommelen et al. 2006; Brunner et al. 2009; Ichinose & Lenoir 2009; Oppelt & Heinze 2009; Peeters & Liebig 2009; Izzo et al. 2010).

Diet. One of the most important factors that influence the CHC phenotype is the diet insects are feeding upon. Several studies of social insects show that their diet affects their CHC profiles (e.g. Liang & Silverman 2000; Buczkowski et al. 2005; Richard et al. 2004; Sorvari et al. 2008; Ichinose et al. 2009; Vonshak et al. 2009; Florane et al. 2004). Furthermore, the CHC profile of adult mustard leaf beetles (*Phaedon cochleariae*) depend on their host plant species. The beetles change their CHC profile after a shift to a novel host plant species; they use fatty acids of their food as precursors for the biosynthesis of their CHCs (Geiselhardt et al. 2012; Otte et al. 2015). Several other insect species have been shown to use diet-derived fatty acids as precursors for their CHC biosynthesis (Blomquist & Jackson 1973; Pennanec'h et al. 1997; Stennett and Etges 1997; Rundle et al. 2005; Blaul & Ruther 2011). The triacylglycerol composition in the diet is known to significantly affect CHC profiles in *D. mojavensis* (Etges et al. 2006). The exact mechanism by which the diet affects the CHC pattern of insects is not yet fully understood. Jurenka (2004) suggests that polyunsaturated cuticular hydrocarbons of moths are synthesized from linoleic or linolenic acid, i.e. compounds that are obtained from the insect's diet. Stingless bees can use resin or resin-derived compounds to enrich their CHC profile (Leonhardt et al. 2011).

5.4 Dynamics of phenotypic change

CHC profiles are dynamic and may change over time. The changes may be of quantitative nature when quantities of CHC compounds vary and thus, the quantitative composition is changed (e.g. Vander Meer et al. 1989; Nielsen et al. 1999; Thurin & Aron 2008), but also qualitative changes over time are observed when novel CHC compounds are added to a CHC pattern (e.g. Boulay et al. 2000; Steiner et al. 2007).

A major question regarding the dynamics of CHC phenotype formation is that of the speed by which the change takes place. How fast do new insect CHC phenotypes develop in a changing environment? The following studies summarize information on the time that is necessary for a change in the CHC profile.

In ants, changes in CHC profiles due to switches from nestwork to foraging resulted in changes of CHC phenotypes over periods ranging from about two weeks (Provost et al. 1993) to about a month (Kaib et al. 2000; Lahav et al. 2001). When the mustard leaf beetle *Phaedon cochleariae* switches to a novel host plant, it takes about 14 days until its novel CHC profile significantly differs from the former one (Geiselhardt et al. 2012).

When insects are exposed to high temperature and low relative humidity, desiccation can occur (Chown et al. 2011). In this case, organisms have to react fast by changing their CHC phenotype to prevent water loss. Howard et al. (1995) showed that the CHC profile of *Oryzaephilus surinamensis* beetles changes rapidly within 1 day when beetles are stressed by high temperature and the danger of desiccation; the beetles increased the quantity of cuticular monoenes and dienes in response to high temperature, whereas low temperatures caused a decrease in the quantity of monoenes. Also other factors like a change in the fertility status or colony odor can lead to a rapid change (< 20 h) in the chemical composition of CHCs (Soroker et al. 1995; Cuvillier-Hot et al. 2005). Mating is also known to trigger immediate changes, e.g. in the ant *Leptothorax gredleri* the CHC profile is almost modified immediately (Oppelt & Heinze 2009). Furthermore, immune stimulation of insects may elicit rapid modifications of their CHC profiles (Richard et al. 2008; Holman et al. 2010; Bos et al. 2011; Nielsen & Holman 2012).

Hence, changes of CHC profiles may take minutes to weeks. So far, it is difficult to predict which type of environmental stimulus induces a slow change of the CHC profile and which one triggers a highly rapid change. The dynamics of a change in an insect's CHC pattern will certainly depend on the mechanisms underlying these changes.

Rapid changes are expected when an insect obtains CHCs from other organisms and/or the environmental substrate by physical contact (e.g. Weddle et al. 2013). Fast changes in the CHC pattern may still occur, when the changes are due to changes in the dynamics of the transport of CHCs to the cuticle via lipophorins (Schal et al. 1998). If the change in an insect's CHC pattern requires a change in CHC biosynthesis, the dynamics of this change will depend on the insect's physiological state (e.g. the dynamics of changes in hormone levels), the availability of CHC precursors and the uptake of compounds affecting biosynthesis activity. If the uptake of novel dietary compounds induces a change of the CHC profile, this change may take several days until it results in a pattern that significantly differs from the original one (e.g. Otte et al. 2015). On the one hand, a shift of an insect to a novel diet may result in a novel pattern of precursors (mainly fatty acids, but also amino acids) for CHC biosynthesis (Howard & Blomquist 2005; Blomquist 2010). On the other hand, an insect feeding on a novel diet may also take up dietary compounds that affect the activity of enzymes that are necessary for CHC biosynthesis (fatty acid synthase, desaturases, and fatty acyl-CoA elongase). In addition, a novel diet may lead to a change in the microbiome of the insect's gut and thus, affect ingestion of dietary compounds that in turn may impact the CHC profile (Sharon et al. 2010). It is known from studies of humans that a dietary switch changes the gut microbiome within a single day (e.g. David et al. 2014).

5.5 Modes of phenotypic change

Exposure to different environmental conditions is suggested to result in two different modes of phenotypic changes:

(A) stepwise (saltatory) shifts between different environments (e.g. colonization of different host plants) with no intermediates or

(B) a continuous functional relationship (gradual) between phenotype and environment (e.g. exposure to gradually changing temperatures) with many intermediates (Figure 5.1; West-Eberhard 2003).

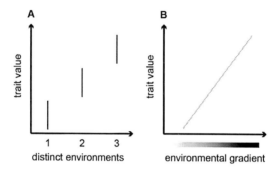

Figure 5.1: Scheme of two modes of phenotypic plasticity: (A) stepwise shifts between different environments (e.g. colonization of different host plants) with no intermediate phenotypes (black lines) and (B) a continuous gradient between phenotype and environment (e.g. temperature) with many intermediate phenotypes (grey line).

The first type of plasticity (A) is characterized by the change in the mean phenotype which is produced in (at least) two different environments. For example, two beetle populations that are feeding for a certain time on different host plants can express qualitatively and quantitatively different CHC phenotypes that might even promote sexual isolation (unpublished data). Between these two environments no intermediate CHC phenotype can develop. The type-(A)-phenotypic plasticity may be quantified by the mean and variance of phenotypes within a population. A shift in a mean phenotype can occur when all individuals respond similarly to an environmental cue. The variance can increase if an individual genotype within a population responds differently to the same cue. Hence, type-(A)-phenotypes within a certain environment can show slight differences (but are mostly similar), but type-(A)-phenotypes occurring in two different environments are clearly different and separable.

The second mode of phenotypic plasticity (B) focuses on a functional relationship between environmental conditions and phenotypes. Here organisms that are exposed to a continuously changing environmental gradient (e.g. increasing temperatures) express continuously different phenotypes (many intermediate phenotypes between different environments). For example, the chain length of

CHCs may increase as the temperatures increase to provide the organisms with protection against water loss (Woodrow et al. 2000).

These plastic responses of either plasticity type outlined above can be developmentally fixed (Price 2006) or rapidly reversible. With respect to insect CHC plasticity, reversibility of CHC phenotypes may help to avoid mismatches between phenotype and changing environment.

5.6 Consequences of phenotypic change

As outlined above, a wide range of factors can influence the quantitative and qualitative chemical composition of an insect CHC phenotype (Via et al. 1995; Pigliucci et al. 2006; Ghalambor et al. 2007; Pfennig et al. 2010; Moczek et al. 2011). What are the benefits and costs of this phenotypic plasticity? When is it adaptive?

An insect benefits from its phenotypic CHC plasticity when a certain phenotype induced by a specific environment has higher fitness in that environment than alternative phenotypes (DeWitt et al. 1998; Snell-Rood et al. 2010). A CHC phenotype is expected to show higher fitness

➢ if it has improved abilities to cope with the conditions in this specific environment. As outlined above, many insects can change their CHC phenotype in response to certain abiotic and biotic conditions and thus, improve their abilities to cope with this environment.

➢ if the costs for production and maintenance of the phenotype do not outweigh the benefit (Reylea 2002; Pigliucci 2005). Costs for reproduction and maintenance of a phenotype may especially occur when a novel CHC phenotype is not just obtained by physical contact, but by biosynthesis of additional or new CHCs (compare section "Dynamics of phenotypic change").

➢ if the dynamics of a change in a CHC profile is fast enough to follow the dynamics of environmental changes inducing a CHC profile. The dynamics of phenotypic changes may range from minutes to weeks. If, for example, a temperature-triggered change in a CHC profile would need a week, the insect might benefit from this change if this novel phenotype would show increased reproductive success at these temperature conditions; however, if the temperature would change significantly every 3 to 4 days, temperature-dependent phenotypic changes that need a week until being established cannot result in higher fitness.

➢ if the phenotype-caused improvement to cope with certain conditions is not outweighed by ecological costs. Which type of ecological costs might an insect need to "pay" when displaying phenotypic plasticity? For example, if an insect changes its CHC profile in response to decreasing temperatures, the novel profile might be more susceptible to pathogen attack or counteract chemical mimicry or impair intraspecific interactions.

If an insect uses its CHCs for mediation of intraspecific interactions, high phenotypic plasticity of the CHC profile is indeed a challenge for these insects, since intraspecific chemical communication requires high reliability of the signals. If signals are not reliable, the receivers will be unable to gain fitness benefits. Moreover, if a novel CHC phenotype cannot be perceived accurately, the behavioral system (signal-perceiver system) will collapse.

Thus, how can insects successfully use CHCs for intraspecific communication, although these compounds are phenotypically so plastic?

> First, when insects change their behaviorally relevant CHC profiles due to different environmental conditions, these changes might be changes in compounds not relevant in intraspecific communication.

> Second, when the alteration of a CHC phenotype is a change in quantities of all behaviorally relevant CHCs, their ratios may stay the same. Here quantities might change, but possibly information conveying ratios of compounds might be kept the same. Ratios of compounds are well known to play a role in numerous chemically mediated intraspecific interactions (Geiselhardt et al. 2012; Weiss et al. 2013).

> Finally, if qualitative and quantitative changes in the chemical composition of a CHC profile occur, a mechanism is required to recognize the new CHC phenotype. The insect may recognize a conspecific individual by comparing its own (learned) chemical phenotype with the profile of the counterpart ("self-referent phenotype matching" Lihoreau & Rivault 2009; Mateo 2010; Weddle et al. 2013). Assuming that insects use self-referent phenotype matching for recognition of conspecifics, the organisms exposed to the same environmental conditions match the chemical template more exactly than individuals from a different environment expressing an alternative CHC phenotype. This process corresponds with the signal matching process of the sensory drive hypothesis (Boughman 2002; Smadja & Butlin 2009). Thus, "self-referent phenotype matching" might help to cope with variable CHC phenotypes if these are important of intraspecific communication, but will also promote divergence in behavioral interactions.

If divergence of CHC profiles that affect sexual behavior leads to assortative mating, this might promote sexual isolation. For example, studies of the mustard leaf beetle *Phaedon cochleariae* reveal that these beetles use their CHC profiles for mate recognition. They prefer partners with similar CHC profiles; the beetles show similar profiles when feeding on the same host plant species (Geiselhardt et al. 2009, 2012; unpublished data). If the beetles show fidelity to a certain host species, their mating preferences for similar CHC phenotypes feeding on the same plant species is expected to lead to genetic divergence in a population and thus, to promote ecological speciation. Similarly, diet-induced changes in the CHC profile of *Drosophila melanogaster* resulted in changes of mating preferences (Rundle et al. 2005; Etges et al. 2006; Havens & Etges 2013). Hence, in these cases phenotypic divergence might promote genetic divergence, and thus finally impact insect speciation.

5.7 Conclusion

A plethora of studies demonstrated the immense plasticity of insect CHC profiles. Nevertheless, further studies are needed on the question why insects use so highly variable chemical signals for intraspecific communication although signal reliability is needed. These studies will provide a deeper understanding on the evolution of mating signals, the respective recognition systems as well as on ecological speciation processes and the impact of phenotype divergence on genetic divergence. Furthermore, even though much knowledge is available on how insects' biosynthesize their CHCs, still many questions remain to be answered on how environmental, especially nutritionally factors influence the biosynthesis of CHCs and its dynamics.

Phenotypic plasticity of insect CHC profiles is favored in changing environments because it allows an individual to adapt its phenotype to novel environmental conditions. Furthermore, plastic CHC phenotypes allow organisms to invade multiple, disparate ecological niches, thus extending the geographic range and decreasing the probability of extinction caused by habitat loss or environmental stochasticity (Snell-Rood et al. 2010). Moreover, if insects use their CHCs for intraspecific communication, phenotypic changes may promote divergence of a population, and if the diverging phenotypes are maintained, finally promote ecological speciation.

Hence, phenotypic plasticity of insect CHC profiles may greatly impact on the fitness of an insect species, its ecological niche and geographical distribution, and thus on the diversity of insect species that evolve.

5.8 References

Agrawal A.A. (2001). Ecology - Phenotypic plasticity in the interactions and evolution of species. *Science*, 294, 321-326.

Akino T., Knapp J.J., Thomas J.A. & Elmes G.W. (1999). Chemical mimicry and host specificity in the butterfly *Maculinea rebeli*, a social parasite of *Myrmica* ant colonies. *Proceedings of the Royal Society B: Biological Sciences*, 266, 1419-1426.

Armold M.T. & Regnier F.E. (1975). Developmental study of cuticular hydrocarbons of *Sarcophaga bullata*. *Journal of Insect Physiology*, 21, 1827-1833.

Beye M., Neumann P., Chapuisat M., Pamilo P. & Moritz R.F.A. (1998). Nestmate recognition and the genetic relatedness of nests in the ant *Formica pratensis*. *Behavioral Ecology and Sociobiology*, 43, 67-72.

Blaul B. & Ruther J. (2011). How parasitoid females produce sexy sons: a causal link between oviposition preference, dietary lipids and mate choice in *Nasonia*. *Proceedings of the Royal Society B: Biological Sciences*, 278, 3286-3293.

Blomquist G.J. (2010). Biosynthesis of cuticular hydrocarbons, pp. 35–52, in G.J. Blomquist and A.G. Bagnères (eds.). Insect Hydrocarbons: Biology, Biochemistry, and Chemical Ecology. *Cambridge University Press, Cambridge*.

Blomquist G.J. & Bagnères, A.-G. (2010) Introduction: history and overview of insect hydrocarbons, pp. 3–18, in G.J. Blomquist and A.G. Bagnères (eds.). Insect Hydrocarbons: Biology, Biochemistry, and Chemical Ecology. *Cambridge University Press, Cambridge*.

Blomquist G.J. & Jackson L.L. (1973). Incorporation of labeled dietary alkanes into cuticular lipids of grasshopper *Melanoplus sanguinipes*. *Journal of Insect Physiology*, 19, 1639-1647.

Boughman J.W. (2002). How sensory drive can promote speciation. *Trends in Ecology & Evolution*, 17, 571-577.

Bonavita-Cougourdan A., Clement J.L. & Lange C. (1987). Nestmate recognition - the role of cuticular hydrocarbons in the ant *Camponotus vagus* Scop. *Journal of Entomological Science*, 22, 1-10.

Bos N., Grinsted L. & Holman L. (2011). Wax on, wax off: nest soil facilitates indirect transfer of recognition cues between ant nestmates. *PLoS ONE*, 6, e19435.

Boulay R., Hefetz A., Soroker V. & Lenoir A. (2000). *Camponotus fellah* colony integration: worker individuality necessitates frequent hydrocarbon exchanges. *Animal Behaviour*, 59, 1127-1133.

Brandt M., van Wilgenburg E., Sulc R., Shea K.J. & Tsutsui N.D. (2009). The scent of supercolonies: the discovery, synthesis and behavioural verification of ant colony recognition cues. *BMC Biology*, 7, 71.

Brunner E., Kroiss J. & Heinze J. (2009). Chemical correlates of reproduction and worker policing in a myrmicine ant. *Journal of Insect Physiology*, 55, 19-26.

Buczkowski G., Kumar R., Suib S. & Silverman J. (2005). Diet-related modification of cuticular hydrocarbon profiles of the Argentine ant, *Linepithema humile*, diminishes intercolony aggression. *Journal of Chemical Ecology*, 31, 829-843.

Chown S.L., Sorensen J.G. & Terblanche J.S. (2011). Water loss in insects: An environmental change perspective. *Journal of Insect Physiology*, 57, 1070-1084.

Coyne J.A. & Orr, H.A. (2004). Speciation. *Sinauer, Sunderland.*

Cuvillier-Hot V., Cobb M., Malosse C. & Peeters C. (2001). Sex, age and ovarian activity affect cuticular hydrocarbons in *Diacamma ceylonense*, a queenless ant. *Journal of Insect Physiology*, 47, 485-493.

Cuvillier-Hot V., Renault V. & Peeters C. (2005). Rapid modification in the olfactory signal of ants following a change in reproductive status. *Naturwissenschaften*, 92, 73-77.

Dani F., Jones G., Corsi S., Beard R., Pradella D. & Turillazzi S. (2005). Nestmate recognition cues in the honey bee: Differential importance of cuticular alkanes and alkenes. *Chemical Senses*, 30, 477-489.

David L.A., et al. (2014). Diet rapidly and reproducibly alters the human gut microbiome. *Nature*, 505, 559–563.

de Biseau J.C., Passera L., Daloze D. & Aron S. (2004). Ovarian activity correlates with extreme changes in cuticular hydrocarbon profile in the highly polygynous ant, *Linepithema humile*. *Journal of Insect Physiology*, 50, 585-593.

Dettner K. & Liepert C. (1994). Chemical mimicry and camouflage. *Annual Review of Entomology*, 39, 129-154.

DeWitt T.J., Sih A. & Wilson D.S. (1998). Costs and limits of phenotypic plasticity. *Trends in Ecology & Evolution*, 13, 77-81.

Drès M. & Mallet J. (2002). Host races in plant-feeding insects and their importance in sympatric speciation. *Philosophical Transactions of the Royal Society of London Series B: Biological Sciences*, 357, 471-492.

Elmes G.W., Akino T., Thomas J.A., Clarke R.T. & Knapp J.J. (2002). Interspecific differences in cuticular hydrocarbon profiles of *Myrmica* ants are sufficiently consistent to explain host specificity by *Maculinea* (large blue) butterflies. *Oecologia*, 130, 525-535.

Espelie K.E. & Payne J.A. (1991). Characterization of the cuticular lipids of the larvae and adults of the pecan weevil, *Curculio caryae*. *Biochemical Systematics and Ecology*, 19, 127-132.

Etges W.J., Veenstra C.L. & Jackson L.L. (2006). Premating isolation is determined by larval rearing substrates in cactophilic *Drosophila mojavensis*. VII. Effects of larval dietary fatty acids on adult epicuticular hydrocarbons. *Journal of Chemical Ecology*, 32, 2629-2646.

Ferveur J. (2005). Cuticular hydrocarbons: their evolution and roles in *Drosophila* pheromonal communication. *Behavior Genetics*, 35, 279-295.

Florane C.B., Bland J.M., Husseneder C. & Raina A.K. (2004). Diet-mediated inter-colonial aggression in the Formosan subterranean termite *Coptotermes formosanus*. *Journal of Chemical Ecology*, 30, 2559-2574.

Geiselhardt S., Geiselhardt S. & Peschke K. (2006). Chemical mimicry of cuticular hydrocarbons - how does *Eremostibes opacus* gain access to breeding burrows of its host *Parastizopus armaticeps* (Coleoptera: Tenebrionidae)? *Chemoecology*, 16, 59-68.

Geiselhardt S., Otte T. & Hilker M. (2009). The role of cuticular Hydrocarbons in male mating behavior of the mustard leaf beetle, *Phaedon cochleariae* (F.). *Journal of Chemical Ecology*, 35, 1162-1171.

Geiselhardt S., Otte T. & Hilker M. (2012). Looking for a similar partner: host plants shape mating preferences of herbivorous insects by altering their contact pheromones. *Ecology Letters*, 15, 971-977.

Ghalambor C.K., McKay J.K., Carroll S.P. & Reznick D.N. (2007). Adaptive versus non-adaptive phenotypic plasticity and the potential for contemporary adaptation in new environments. *Functional Ecology*, 21, 394-407.

Gibbs A. (1998). Water-proofing properties of cuticular lipids. *American Zoologist*, 38, 471-482.

Gibbs A., Kuenzli M. & Blomquist G. (1995). Sex-related and age-related changes in the biophysical properties of cuticular lipids of the housefly, *Musca domestica*. *Archives of Insect Biochemistry and Physiology*, 29, 87-97.

Gibbs A.G. (2002). Lipid melting and cuticular permeability: new insights into an old problem. *Journal of Insect Physiology*, 48, 391-400.

Gibbs A.G., Louie A.K. & Ayala J.A. (1998). Effects of temperature on cuticular lipids and water balance in a desert *Drosophila*: Is thermal acclimation beneficial? *Journal of Experimental Biology*, 201, 71-80.

Gibbs A.G. & Rajpurohit S. (2010). Water-proofing properties of cuticular lipids, pp. 100-119, in G.J. Blomquist and A.G. Bagnères (eds.). Insect Hydrocarbons: Biology, Biochemistry, and Chemical Ecology. *Cambridge University Press, Cambridge*.

Girotti J.R., Mijailovsky S.J. & Patricia Juarez M. (2012). Epicuticular hydrocarbons of the sugarcane borer *Diatraea saccharalis* (Lepidoptera: Crambidae). *Physiological Entomology*, 37, 266-277.

Goodrich B.S. (1970). Cuticular lipids of adults and puparia of Australian sheep blowfly *Lucilia cuprina* (Wied). *Journal of Lipid Research*, 11, 1-6.

Hadley N.F. (1977). Epicuticular lipids of desert tenebrionid beetle, *Eleodes armata* - seasonal and acclimatory effects on composition. *Insect Biochemistry*, 7, 277-283.

Havens J.A. & Etges W.J. (2013). Premating isolation is determined by larval rearing substrates in cactophilic *Drosophila mojavensis*. IX. Host plant and population specific epicuticular hydrocarbon expression influences mate choice and sexual selection. *Journal of Evolutionary Biology*, 26, 562-576.

Holman L., Jorgensen C.G., Nielsen J. & d'Ettorre P. (2010). Identification of an ant queen pheromone regulating worker sterility. *Proceedings of the Royal Society B: Biological Sciences*, 277, 3793-3800.

Howard R.W. (1993). Cuticular hydrocarbons and chemical communication, pp. 179-226, in D.W. Stanley-Samuelson and D.R. Nelson (eds.). Insect Lipids. Chemistry, Biochemistry and Biology. *University of Nebraska Press, Lincoln*.

Howard R.W. & Blomquist G. (2005). Ecological, behavioral, andbiochemical aspects of insect hydrocarbons. *Annual Review of Entomology*, 50, 371-393.

Howard R.W., Howard C.D. & Colquhoun S. (1995). Ontogenic and environmentally-induced changes in cuticular hydrocarbons of *Oryzaephilus surinamensis* (Coleoptera: Cucujidae). *Annals of the Entomological Society of America*, 88, 485-495.

Hugo L.E., Kay B.H., Eaglesham G.K., Holling N. & Ryan P.A. (2006). Investigation of cuticular hydrocarbons for determining the age and survivorship of Australasian mosquitoes. *American Journal of Tropical Medicine and Hygiene*, 74, 462-474.

Ichinose K., Boulay R., Cerda X. & Lenoir A. (2009). Influence of queen and diet on nestmate recognition and cuticular hydrocarbon differentiation in a fission-dispersing ant, *Aphaenogaster senilis*. *Zoological Science*, 26, 681-685.

Ichinose K. & Lenoir A. (2009). Ontogeny of hydrocarbon profiles in the ant *Aphaenogaster senilis* and effects of social isolation. *Comptes Rendus Biologies*, 332, 697-703.

Izzo A., Wells M., Huang Z. & Tibbetts E. (2010). Cuticular hydrocarbons correlate with fertility, not dominance, in a paper wasp, *Polistes dominulus*. *Behavioral Ecology and Sociobiology*, 64, 857-864.

Johnson C.A., Vander Meer R.K. & Lavine B. (2001). Changes in the cuticular hydrocarbon profile of the slave-maker ant queen, *Polyergus breviceps* Emery, after killing a *Formica* host queen (Hymenoptera: Formicidae). *Journal of Chemical Ecology*, 27, 1787-1804.

Jurenka R. (2004). Insect pheromone biosynthesis. *Topics in Current Chemistry*, 239, 97–131.

Kaib M., Eisermann B., Schoeters E., Billen J., Franke S. & Francke W. (2000). Task-related variation of postpharyngeal and cuticular hydrocarhon compositions in the ant *Myrmicaria eumenoides*. *Journal of Comparative Physiology A: Sensory, Neural, and Behavioral Physiology*, 186, 939-948.

Kuo T.H., Yew J.Y., Fedina T.Y., Dreisewerd K., Dierick H.A. & Pletcher S.D. (2012). Aging modulates cuticular hydrocarbons and sexual attractiveness in *Drosophila melanogaster*. *Journal of Experimental Biology*, 215, 814-821.

Lahav S., Soroker V., Vander Meer R.K. & Hefetz A. (2001). Segregation of colony odor in the desert ant *Cataglyphis niger*. *Journal of Chemical Ecology*, 27, 927-943.

Lecuona R., Riba G., Cassier P. & Clement J.L. (1991). Alterations of insect epicuticular hydrocarbons during infection with *Beauveria bassiana* or *B. brongniartii*. *Journal of Invertebrate Pathology*, 58, 10-18.

Lenoir A., D'Ettorre P., Errard C. & Hefetz A. (2001). Chemical ecology and social parasitism in ants. *Annual Review of Entomology*, 46, 573-599.

Leonhardt S.D., Wallace H.M. & Schmitt T. (2011). The cuticular profiles of Australian stingless bees are shaped by resin of the eucalypt tree *Corymbia torelliana*. *Austral Ecology*, 36, 537-543.

Liang D. & Silverman J. (2000). "You are what you eat": Diet modifies cuticular hydrocarbons and nestmate recognition in the Argentine ant, *Linepithema humile*. *Naturwissenschaften*, 87, 412-416.

Liebig J. (2010). Hydrocarbon profiles indicate fertility and dominance status in ant, bee, and wasp colonies, pp. 245–281, in G.J. Blomquist and A.G. Bagnères (eds.). Insect Hydrocarbons: Biology, Biochemistry, and Chemical Ecology. *Cambridge University Press, Cambridge.*

Liebig J., Peeters C., Oldham N.J., Markstadter C. & Holldobler B. (2000). Are variations in cuticular hydrocarbons of queens and workers a reliable signal of fertility in the ant *Harpegnathos saltator*? *Proceedings of the National Academy of Sciences of the United States of America*, 97, 4124-4131.

Lihoreau M. & Rivault C. (2009). Kin recognition via cuticular hydrocarbons shapes cockroach social life. *Behavioral Ecology*, 20, 46-53.

Lommelen E., Johnson C.A., Drijfhout F.P., Billen J., Wenseleers T. & Gobin B. (2006). Cuticular hydrocarbons provide reliable cues of fertility in the ant *Gnamptogenys striatula*. *Journal of Chemical Ecology*, 32, 2023-2034.

Lucas C., Pho D.B., Jallon J.M. & Fresneau D. (2005). Role of cuticular hydrocarbons in the chemical recognition between ant species in the *Pachycondyla villosa* species complex. *Journal of Insect Physiology*, 51, 1148-1157.

Martin S. & Drijfhout F. (2009). A review of ant cuticular hydrocarbons. *Journal of Chemical Ecology*, 35, 1151-1161.

Mateo J.M. (2010). Self-referent phenotype matching and long-term maintenance of kin recognition. *Animal Behaviour*, 80, 929-935.

Miner B.G., Sultan S.E., Morgan S.G., Padilla D.K. & Relyea R.A. (2005). Ecological consequences of phenotypic plasticity. *Trends in Ecology & Evolution*, 20, 685-692.

Moczek A.P. (2010). Phenotypic plasticity and diversity in insects. *Philosophical Transactions of the Royal Society B: Biological Sciences*, 365, 593-603.

Moczek A.P., Sultan S., Foster S., Ledon-Rettig C., Dworkin I., Nijhout H.F., Abouheif E. & Pfennig D.W. (2011). The role of developmental plasticity in evolutionary innovation. *Proceedings of the Royal Society B: Biological Sciences*, 278, 2705-2713.

Mpuru S., Blomquist G.J., Schal C., Roux M., Kuenzli M., Dusticier G., Clement J.L. & Bagnères A.G. (2001). Effect of age and sex on the production of internal and external hydrocarbons and pheromones in the housefly, *Musca domestica*. *Insect Biochemistry and Molecular Biology*, 31, 139-155.

Nelson D.R. & Lee R.E. (2004). Cuticular lipids and desiccation resistance in overwintering larvae of the goldenrod gall fly, *Eurosta solidaginis* (Diptera: Tephritidae). *Comparative Biochemistry and Physiology - Part B: Biochemistry & Molecular Biology*, 138, 313-320.

Nielsen J., Boomsma J.J., Oldham N.J., Petersen H.C. & Morgan E.D. (1999). Colony-level and season-specific variation in cuticular hydrocarbon profiles of individual workers in the ant *Formica truncorum*. *Insectes Sociaux*, 46, 58-65.

Nielsen M.L. & Holman L. (2012). Terminal investment in multiple sexual signals: immune-challenged males produce more attractive pheromones. *Functional Ecology*, 26, 20-28.

Noorman N. & Den Otter C.J. (2002). Effects of relative humidity, temperature, and population density on production of cuticular hydrocarbons in housefly *Musca domestica* L. *Journal of Chemical Ecology*, 28, 1819-1829.

Oppelt A. & Heinze J. (2009). Mating is associated with immediate changes of the hydrocarbon profile of *Leptothorax gredleri* ant queens. *Journal of Insect Physiology*, 55, 624-628.

Otte T., Hilker M. & Geiselhardt S. (2015). The effect of dietary fatty acids on the cuticular hydrocarbon phenotype of an herbivorous insect and consequences for mate recognition. *Journal of Chemical Ecology*, 41, 32-43.

Pedrini N., Ortiz-Urquiza A., Huarte-Bonnet C., Zhang S. & Keyhani N.O. (2013). Targeting of insect epicuticular lipids by the entomopathogenic fungus *Beauveria bassiana*: hydrocarbon oxidation within the context of a host-pathogen interaction. *Frontiers in Microbiology*, 4, 24.

Peeters C., Monnin T. & Malosse C. (1999). Cuticular hydrocarbons correlated with reproductive status in a queenless ant. *Proceedings of the Royal Society B: Biological Sciences*, 266, 1323-1327.

Peeters C. & Liebig J. (2009). Fertility signaling as a general mechanism of regulating reproductive division of labor in ants, pp. 220–242, in J. Gadau and J. Fewell (eds.). Organization of insect societies: from genome to sociocomplexity. *Harvard University Press, Cambridge*.

Pennanec'h M., Bricard L., Kunesch G. & Jallon J.M. (1997). Incorporation of fatty acids into cuticular hydrocarbons of male and female *Drosophila melanogaster*. *Journal of Insect Physiology*, 43, 1111-1116.

Peschke K. (1985). Immature males of *Aleochara curtula* avoid intrasexual aggressions by producing the female sex-pheromone. *Naturwissenschaften*, 72, 274-275.

Peschke K. (1987). Cuticular hydrocarbons regulate mate recognition, male aggression, and female choice of the rove beetle, *Aleochara curtula*. *Journal of Chemical Ecology*, 13, 1993-2008.

Pfennig D.W., Wund M.A., Snell-Rood E.C., Cruickshank T., Schlichting C.D. & Moczek A.P. (2010). Phenotypic plasticity's impacts on diversification and speciation. *Trends in Ecology & Evolution*, 25, 459-467.

Pigliucci M. (2005). Evolution of phenotypic plasticity: where are we going now? *Trends in Ecology & Evolution*, 20, 481-486.

Pigliucci M., Murren C.J. & Schlichting C.D. (2006). Phenotypic plasticity and evolution by genetic assimilation. *Journal of Experimental Biology*, 209, 2362-2367.

Price T.D. (2006). Phenotypic plasticity, sexual selection and the evolution of colour patterns. *Journal of Experimental Biology*, 209, 2368-2376.

Provost E., Riviere G., Roux M., Morgan E.D. & Bagnères A.-G. (1993). Change in the chemical signature of the ant *Leptothorax lichtensteini* Bondroit with time. *Insect Biochemistry and Molecular Biology*, 23, 945-957.

Reylea R.A. (2002). Costs of phenotypic plasticity. *The American Naturalist*, 159, 272–282.

Richard F.J., Aubert A. & Grozinger C.M. (2008). Modulation of social interactions by immune stimulation in honey bee, *Apis mellifera*, workers. *BMC Biology*, 6, 13.

Richard F.J., Hefetz A., Christides J.P. & Errard C. (2004). Food influence on colonial recognition and chemical signature between nestmates in the fungus-growing ant *Acromyrmex subterraneus*. *Chemoecology*, 14, 9-16.

Rouault J.D., Marican C., Wicker-Thomas C. & Jallon J.M. (2004). Relations between cuticular hydrocarbon (HC) polymorphism, resistance against desiccation and breeding temperature; a model for HC evolution in *D. melanogaster* and *D. simulans*. *Genetica*, 120, 195-212.

Roux O., Gers C. & Legal L. (2006). When, during ontogeny, waxes in the blowfly (Calliphoridae) cuticle can act as phylogenetic markers. *Biochemical Systematics and Ecology*, 34, 406-416.

Rundle H.D., Chenoweth S.F., Doughty P. & Blows M.W. (2005). Divergent selection and the evolution of signal traits and mating preferences. *PLoS Biology*, 3, 1988-1995.

Ruther J., Doering M. & Steiner S. (2011). Cuticular hydrocarbons as contact sex pheromone in the parasitoid *Dibrachys cavus*. *Entomologia Experimentalis et Applicata*, 140, 59-68.

Salamin N., Wuest R.O., Lavergne S., Thuiller W. & Pearman P.B. (2010). Assessing rapid evolution in a changing environment. *Trends in Ecology & Evolution*, 25, 692-698.

Salvy M., Martin C., Bagneres A.G., Provost E., Roux M., Le Conte Y. & Clement J.L. (2001). Modifications of the cuticular hydrocarbon profile of *Apis mellifera* worker bees in the presence of the ectoparasitic mite *Varroa jacobsoni* in brood cells. *Parasitology*, 122, 145-159.

Schal C., Sevala V.L., Young H.P. & Bachmann J.A.S. (1998). Sites of synthesis and transport pathways of insect hydrocarbons: Cuticle and ovary as target tissues. *American Zoologist*, 38, 382-393.

Schlichting, C.D. & Pigliucci, M. (1998). Phenotypic evolution: a reaction norm perspective. *Sinauer Associates, Sunderland*.

Scott M.P., Madjid K. & Orians C.M. (2008). Breeding alters cuticular hydrocarbons and mediates partner recognition by burying beetles. *Animal Behaviour*, 76, 507-513.

Sharon G., Segal D., Ringo J.M., Hefetz A., Zilber-Rosenberg I. & Rosenberg E. (2010). Commensal bacteria play a role in mating preference of *Drosophila melanogaster*. *Proceedings of the National Academy of Sciences of the United States of America*, 107, 20051-20056.

Singer T. (1998). Roles of hydrocarbons in the recognition systems of insects. *American Zoologist*, 38, 394-405.

Smadja C. & Butlin R.K. (2009). On the scent of speciation: the chemosensory system and its role in premating isolation. *Heredity*, 102, 77-97.

Snell-Rood E.C., Van Dyken J.D., Cruickshank T., Wade M.J. & Moczek A.P. (2010). Toward a population genetic framework of developmental evolution: the costs, limits, and consequences of phenotypic plasticity. *Bioessays*, 32, 71-81.

Snell-Rood E.C. (2012). Selective Processes in Development: Implications for the Costs and Benefits of Phenotypic Plasticity. *Integrative and Comparative Biology*, 52, 31-42.

Soroker V., Vienne C. & Hefetz A. (1995). Hydrocarbon dynamics within and between nestmates in *Cataglyphis niger* (Hymenoptera: Formicidae). *Journal of Chemical Ecology*, 21, 365-378.

Sorvari J., Theodora P., Turillazzi S., Hakkarainen H. & Sundstrom L. (2008). Food resources, chemical signaling, and nest mate recognition in the ant *Formica aquilonia*. *Behavioral Ecology*, 19, 441-447.

Steiger S., Peschke K., Francke W. & Muller J. (2007). The smell of parents: breeding status influences cuticular hydrocarbon pattern in the burying beetle *Nicrophorus vespilloides*. *Proceedings of the Royal Society B: Biological Sciences*, 274, 2211-2220.

Steiger S., Peschke K. & Muller J.K. (2008). Correlated changes in breeding status and polyunsaturated cuticular hydrocarbons: the chemical basis of nestmate recognition in the burying beetle *Nicrophorus vespilloides*? *Behavioral Ecology and Sociobiology*, 62, 1053-1060.

Steiner S., Mumm R. & Ruther J. (2007). Courtship pheromones in parasitic wasps: Comparison of bioactive and inactive hydrocarbon profiles by multivariate statistical methods. *Journal of Chemical Ecology*, 33, 825-838.

Stennett M.D. & Etges W.J. (1997). Premating isolation is determined by larval rearing substrates in cactophilic *Drosophila mojavensis*. III. Epicuticular hydrocarbon variation is determined by use of different host plants in *Drosophila mojavensis* and *Drosophila arizonae*. *Journal of Chemical Ecology*, 23, 2803-2824.

Thomas M.L. & Simmons L.W. (2011). Short-term phenotypic plasticity in long-chain cuticular hydrocarbons. *Proceedings of the Royal Society B: Biological Sciences*, 278, 3123-3128.

Thurin N. & Aron S. (2008). Seasonal nestmate recognition in the polydomous ant *Plagiolepis pygmaea*. *Animal Behaviour*, 75, 1023-1030.

Tillman J.A., Seybold S.J., Jurenka R.A. & Blomquist G.J. (1999). Insect pheromones - an overview of biosynthesis and endocrine regulation. *Insect Biochemistry and Molecular Biology*, 29, 481-514.

Toolson E.C. (1982). Effects of rearing temperature on cuticle permeability and epicuticular lipid-composition in *Drosophila pseudoobscura*. *Journal of Experimental Zoology*, 222, 249-253.

Trabalon M., Campan M., Clement J.L., Lange C. & Miquel M.T. (1992). Cuticular hydrocarbons of *Calliphora vomitoria* (Diptera) - relation to age and sex. *General and Comparative Endocrinology*, 85, 208-216.

Trabalon M., Plateaux L., Peru L., Bagnères A.G. & Hartmann N. (2000). Modification of morphological characters and cuticular compounds in worker ants *Leptothorax nylanderi* induced by endoparasites *Anomotaenia brevis*. *Journal of Insect Physiology*, 46, 169-178.

Tregenza T., Buckley S., Pritchard V. & Butlin R. (2000). Inter- and intrapopulation effects of sex and age on epicuticular composition of meadow grasshopper, *Chorthippus parallelus*. *Journal of Chemical Ecology*, 26, 257-278.

van Wilgenburg E., Symonds M.R.E. & Elgar M.A. (2011). Evolution of cuticular hydrocarbon diversity in ants. *Journal of Evolutionary Biology*, 24, 1188-1198.

van Zweden J.S., Dreier S. & d'Ettorre P. (2009). Disentangling environmental and heritable nestmate recognition cues in a carpenter ant. *Journal of Insect Physiology*, 55, 158-163.

Vander Meer R.K., Saliwanchik D. & Lavine B. (1989). Temporal changes in colony cuticular hydrocarbon patterns of *Solenopsis invicta* implications for nestmate recognition. *Journal of Chemical Ecology*, 15, 2115-2125.

Vander Meer R.K. & Wojcik D.P. (1982). Chemical mimicry in the myrmecophilous beetle *Myrmecaphodius excavaticollis*. *Science*, 218, 806-808.

Via S., Gomulkiewicz R., Dejong G., Scheiner S.M., Schlichting C.D. & Vantienderen P.H. (1995). Adaptive phenotypic plasticity - consensus and controversy. *Trends in Ecology & Evolution*, 10, 212-217.

Vonshak M., Dayan T., Foucaud J., Estoup A. & Hefetz A. (2009). The interplay between genetic and environmental effects on colony insularity in the clonal invasive little fire ant *Wasmannia auropunctata*. *Behavioral Ecology and Sociobiology*, 63, 1667-1677.

Wagner D., Tissot M. & Gordon D. (2001). Task-related environment alters the cuticular hydrocarbon composition of harvester ants. *Journal of Chemical Ecology*, 27, 1805-1819.

Weddle C.B., Steiger S., Hamaker C.G., Ower G.D., Mitchell C., Sakaluk S.K. & Hunt J. (2013). Cuticular hydrocarbons as a basis for chemosensory self-referencing in crickets: a potentially universal mechanism facilitating polyandry in insects. *Ecology Letters*, 16, 346-53.

Weiss I., Rössler T., Hofferberth J., Brummer M., Ruther J. & Stökl J. (2013). A nonspecific defensive compound evolves into a competition-avoidance cue and a female sex-pheromone. *Nature Communications*, 4, 2767.

West-Eberhard M.J. (2003). Developmental plasticity and evolution. *Oxford University Press, Oxford.*

Woodrow R.J., Grace J.K., Nelson L.J. & Haverty M.I. (2000). Modification of cuticular hydrocarbons of *Cryptotermes brevis* (Isoptera: Kalotermitidae) in response to temperature and relative humidity. *Environmental Entomology*, 29, 1100-1107.

Wund M.A. (2012). Assessing the impacts of phenotypic plasticity on evolution. *Integrative and Comparative Biology*, 52, 5-15.

Zhu G.H., Ye G.Y., Hu C., Xu X.H. & Li K. (2006). Development changes of cuticular hydrocarbons in *Chrysomya rufifacies* larvae: potential for determining larval age. *Medical and Veterinary Entomology*, 20, 438-444.

Summary

The general objective of this PhD thesis is to understand the causes and consequences of phenotypic plasticity of mating recognition systems within and between populations of herbivorous insects. The thesis elucidates the role of host plant use as a key component in ecological speciation processes that promote assortative mating of insects by phenotypically changing their mating signals, thus leading to sexual isolation.

The studies were conducted with the mustard leaf beetles *Phaedon cochleariae* and *P. armoraciae*, feeding either on brassicaceous plants or on Plantaginaceae. The main emphasis of this thesis was placed on

(i) the chemical characterization of the sexual communication system of the beetles (cuticular hydrocarbons = **CHC**s = mate recognition cues),
(ii) the question if the patterns of the identified CHCs are phenotypically plastic and dependent on the host plant a beetle is feeding upon, and
(iii) the question whether host plant-dependent beetle CHC patterns promote assortative mating and thus, lead to sexual isolation.

Gas chromatography coupled with mass spectrometry and behavioral assays were used to address these questions.

Host plants shape mating preferences of herbivorous insects

In a first approach, it was investigated whether divergent host plant use by an herbivorous insect causes assortative mating by phenotypically altering traits involved in mate recognition. To analyze and compare the mating behavior and CHC profiles of *P. cochleariae*, beetles were derived from a common laboratory stock colony on Chinese cabbage (*Brassica rapa* ssp. *pekinensis*) and separated into two groups during the pupal stage. After eclosion, one group of adult beetles was further provided with Chinese cabbage, while the other one was shifted to a novel host plant species and fed with watercress (*Nasturtium officinale*). First, it was tested whether males discriminate behaviorally between "same host plant females" and "different host plant females", i.e. females that fed on another host plant than the males. Mating bioassays clearly showed that male *P. cochleariae* of both groups showed a higher propensity to mate with females that had been fed with the same plant species than with females provided with a different plant species.

In a second step, I chemically compared the mating signals, i.e. CHC profiles of (male and female) adult *P. cochleariae* kept on watercress or Chinese cabbage and investigated how long a beetle needs to feed on a novel plant species until its CHC profile has reached a pattern that is characteristic for the novel host species. The results clearly showed that the beetles' CHC phenotypes were host plant specific and changed within two weeks after a shift to a novel host plant species.

These findings gave rise to the idea that plant-induced phenotypic divergence in mate recognition cues of herbivorous insects may act as an early barrier to gene flow between insect populations on different host species, thus preceding genetic divergence and thus, promoting ecological speciation.

A further approach to elucidating the role of host plants (as ecological factor) in ecological speciation of herbivorous insects was to study the impact of a host shift on phenotypic divergence of two closely related leaf beetles, *P. cochleariae* and *P. armoraciae*. These species co-occur in the same habitat, but are feeding in their natural habitats on different host plant species. In interspecific mating trials, both beetle species showed significant sexual isolation when reared on their natural host plant species, but were lacking sexual isolation when feeding on the same host. This provided first evidence that colonizing a new environment (host plant) may change sexual traits of an herbivorous insect due to the formation of a new chemical insect phenotype which causes changes in mating preferences between sympatric insect species. Subsequent chemical analysis of the CHC profiles of *P. cochleariae* and *P. armoraciae* demonstrated that the host plant species which an insect is feeding upon can significantly change chemical traits important for mate and species recognition. Discrimination of mates depended on the beetles' CHC phenotypes which differed between the beetle species when they fed on the different native host species, but converged when beetles fed upon a common host.

Thus, the behavioral isolation of *P. cochleariae* and *P. armoraciae* and avoidance of sexual interference between them is mediated by host-induced phenotypic plasticity rather than by genetic divergence of their mating signals.

In a final approach, I have addressed the question whether the composition of dietary fatty acids ingested by *P. cochleariae* affects the CHC phenotype of adults, and thus mating preferences. Dietary fatty acids are well known as precursors of insect CHCs. In a first step, the fatty acid composition of two host plant species, i.e. Chinese cabbage and watercress was analyzed. It was investigated how the plant´s fatty acid composition affects the CHC phenotype of the beetles. To scrutinize the impact of dietary fatty acids on the CHC profiles, beetles were reared on three alternative semi-artificial diets differing only in the compositions of fatty acids. The CHC patterns of these beetles were compared chemically. To test whether divergence in CHC phenotypes induced by feeding on the different artificial diets results in diet-specific assortative mating, mating bioassays with all possible host plant dependent male × female combinations were conducted. The study demonstrated that the composition of ingested fatty acids significantly affects the composition of the CHC profile of adult *P. cochleariae*. Feeding on either Chinese cabbage or watercress resulted in different quantitative patterns of both straight-chained and methyl-branched beetle CHCs profiles. Beetles feeding on either of the two host plant species that varied in their fatty acid composition (ratio of mono- to diunsaturated fatty acids) showed specific CHC profiles that differed in ratios of the respective straight-chained hydrocarbons.

These findings strongly suggest that dietary fatty acids are used as precursors for the biosynthesis of straight-chained CHCs in *P. cochleariae*. Divergence of CHC patterns of beetles was detected when beetles fed upon artificial diets which differed only with respect to their fatty acid compositions. Beetles that fed on these different types of artificial diets showed clear differences in their pattern of straight-chained and methyl-branched CHCs. Hence, ingestion of certain blends of fatty acids may significantly determine the CHC pattern of *P. cochleariae*.

Phenotypic plasticity of cuticular hydrocarbon profiles in insects

In addition to the experimental work, current literature on phenotypic plasticity of insect CHC profiles in response to different abiotic and biotic factors was summarized and discussed. Comprehensive knowledge about the environmental effects on insect CHC profiles is essential to understand the mechanisms that underlie phenotypic divergence and the consequences of this divergence. The review considers the "*dynamics of phenotypic change*" by addressing how rapid a change of an insect CHC phenotype may take place; it further analyzes two "*modes of phenotypic change*" and differentiates between gradual and saltational changes. Finally, the review critically discusses the "*consequences of phenotypic change*" in an evolutionary biology context and considers the consequences of plasticity of CHC phenotypes for speciation processes.

—

7

Zusammenfassung

Das Ziel dieser Dissertation ist es, Ursachen und Folgen der phänotypischen Plastizität von Paarungssignalen und Partnerwahl innerhalb und zwischen Populationen von herbivoren Insekten zu analysieren. Die Arbeit soll die Rolle von Wirtspflanzen als Schlüsselkomponente bei ökologischen Artbildungsprozessen verdeutlichen und dabei zeigen, wie die phänotypische Veränderung von Paarungssignalen zu „assortative mating" führen kann und somit sexuelle Isolation fördert.

Die verschiedenen Studien wurden mit den Meerrettichblattkäfern *Phaedon cochleariae* und *P. armoraciae* durchgeführt, welche entweder auf Brassicaceae oder Plantaginaceae gehalten wurden. Dabei wurde der Schwerpunkt dieser Arbeit auf folgende Aspekte gelegt:

(i) die chemische Charakterisierung des sexuellen Kommunikationssystems der beiden Käfer (kutikulare Kohlenwasserstoffe = KKW = Paarungssignale)

(ii) die Frage, ob die Muster der identifizierten KKW der Käfer phänotypisch plastisch und abhängig von der Wirtspflanze sind, auf welchen die Käfer gehalten wurden, sowie

(iii) die Frage, ob wirtspflanzenspezifische KKW der Käfer zu „assortative mating" führen und damit sexuelle Isolation begünstigen.

Für die Untersuchungen wurden methodisch gekoppelte Gaschromatographie-Massenspektrometrie sowie Verhaltenstests eingesetzt.

Wirtspflanzenwechsel bei herbivoren Insekten

In einem ersten Ansatz wurde untersucht, ob divergierende Wirtspflanzennutzung bei herbivoren Insekten zu phänotypischer Veränderung der Paarungssignale führen kann und somit „assortative mating" fördert. Um das Paarungsverhalten sowie die chemische Zusammensetzung der KKW-Profile von *P. cochleariae* zu untersuchen, wurden Käfer aus einer Chinakohl-Laboranzucht während des Puppenstadiums in zwei Gruppen aufgeteilt. Nach dem Schlüpfen der Adulten wurde eine Anzuchtlinie weiterhin auf Chinakohl (*Brassica rapa* ssp. *pekinensis*) gehalten, die zweite Anzuchtlinie wurde mit einer neuen Wirtspflanze (Brunnenkresse; *Nasturtium officinale*) versorgt. Zunächst wurde untersucht, ob männliche Käfer zwischen Weibchen, welche auf der gleichen oder auf einer anderen Fraßpflanze angezogen wurden, unterscheiden können. Die Kopulationsbiotests zeigten dabei deutlich, dass männliche *P. cochleariae* solche Weibchen bevorzugten, die mit derselben Pflanzenart wie sie selbst gefüttert wurden, im Gegensatz zu Weibchen, die auf einer anderen Pflanzenart fraßen.

In einem zweiten Schritt wurden die Paarungssignale chemisch analysiert und verglichen, d.h. KKW-Profile von (männlichen und weiblichen) *P. cochleariae*, welche einerseits auf Brunnenkresse oder auf Chinakohl gehalten wurden. Weiterhin wurde untersucht, wie lange ein Käfer nach einem Wechsel zu einer neuen Wirtspflanze auf dieser fressen muss, bis sein KKW-Profil charakteristisch für diese Wirtspflanze ist. Die Ergebnisse zeigten deutlich, dass die KKW-Phänotypen der Käfer wirtspflanzenspezifisch waren und sich innerhalb von zwei Wochen nach einem Wirtspflanzenwechsel unterschieden.

Diese Ergebnisse gaben Anlass zur Vermutung, dass die durch einen Wirtspflanzenwechsel induzierte phänotypische Divergenz der Paarungssignale als frühe Barriere für den Genfluss zwischen Insektenpopulationen dient und damit die genetische Divergenz fördert und letztlich zur ökologischen Artbildung führen kann.

In einem weiteren Ansatz zur Bestimmung der Rolle von Wirtspflanzen (als ökologischer Faktor) in Artbildungsprozessen wurde untersucht, welche Auswirkung ein Wirtspflanzenwechsel bei nahe verwandten Blattkäfern *P. cochleariae* und *P. armoraciae* hat. Beide Arten teilen sich den gleichen Lebensraum, fressen aber in ihren natürlichen Habitaten auf verschiedenen Wirtspflanzenarten. Interspezifische Verhaltensbiotests zeigten eindeutig sexuelle Isolation, wenn beide Käfer im Labor auf ihrer jeweils natürlichen Wirtspflanze gehalten wurden. Keine sexuelle Isolation wurde beobachtet, wenn beide Arten auf der gleichen Wirtspflanzenart gehalten wurden. Diese Ergebnisse boten erste Hinweise darauf, dass die Kolonisierung einer neuen Umwelt mittels Wirtspflanzenwechsel zur Anpassung/Veränderung der Paarungssignale führen kann und somit die Partnererkennung zwischen sympatrischen Arten beeinflusst. Weiterführende chemische Untersuchungen der KKW-Profile von *P. cochleariae* und *P. armoraciae* zeigten eindeutig, dass die chemischen Käfer-Phänotypen und somit die Signale, welche für Art- und Partnererkennung wichtig sind, vom Fraß an der jeweiligen Wirtspflanzenart abhängig waren. Entsprechend war auch die Verhaltensisolation der Paarungspartner vom KKW-Profil der Käfer abhängig. Konsumierten beide Käferarten verschiedene Wirtspflanzenarten, unterschieden sich ihre KKW-Profile, fraßen sie jedoch die gleiche Wirtspflanzenart, wiesen die Kohlenwasserstoff-Profile der Käfer große Ähnlichkeiten auf.

Folglich wurde die Verhaltensisolation bei *P. cochleariae* und *P. armoraciae* und die Vermeidung von sexueller Interferenz durch wirtspflanzenspezifische phänotypische Plastizität und nicht durch genetische Divergenz der Paarungssignale begünstigt.

In einem letzten Ansatz wurde der Frage nachgegangen, ob das Fettsäuremuster der Nahrung einen Einfluss auf das Kohlenwasserstoff-Profil hat und somit Paarungspräferenzen begünstigen. Aus der Literatur ist bekannt, dass Fettsäuren, welche durch die Nahrung aufgenommen werden, als Vorstufen für die Kohlenwasserstoff-Biosynthese bei Insekten genutzt werden können. In einem ersten Schritt wurde die Fettsäurezusammensetzung der beiden Wirtspflanzen Chinakohl und Brunnenkresse analysiert und untersucht, ob die pflanzlichen Fettsäuren das KKW-Profil der Käfer beeinflussen. Weiterhin wurde anhand von drei verschiedenen Typen von künstlichem Futter mittels chemischer Analysen untersucht, ob unterschiedliche Fettsäurezusammen-setzungen die KKW-Profile der Käfer beeinflussen. In einem nächsten Schritt wurde durch Verhaltensbiotests mit allen möglichen Kombinationen untersucht, ob die Veränderung des KKW-Profils - bedingt durch die drei verschiedenen Fettsäurezusammensetzungen des Kunstfutters - zu nahrungsspezifischem Paarungsverhalten führt. Die Ergebnisse zeigten eindeutig, dass unterschiedliche Fettsäuremischungen im Futter zu unterschiedlichen KKW-Profilen adulter *P. cochleariae* führen. Das KKW-Profil von Käfern, welche entweder auf Chinakohl oder Brunnenkresse gehalten wurden, unterschied sich eindeutig in der quantitativen chemischen Zusammensetzung der geradkettigen und methyl-verzweigten Kohlenwasserstoffe.

Diese Ergebnisse gaben Anlass zur Vermutung, dass Fettsäuren, welche durch die Nahrung aufgenommen werden, als Vorstufen für die Biosynthese von geradkettigen Kohlenwasserstoffen bei *P. cochleariae* verwendet werden. Eine eindeutige Divergenz bei den KKW-Profilen der Käfer konnte festgestellt werden, wenn diese auf künstlichem Futter gehalten wurden, welches sich nur durch die Fettsäurezusammensetzung unterscheidet. Käfer, die auf den drei verschiedenen Typen des künstlichen Futters fraßen, zeigten eindeutige Unterschiede in der chemischen Zusammensetzung der geradkettigen und methylverzweigten Kohlenwasserstoffe. Somit kann die orale Aufnahme von bestimmen Fettsäuren bei *P. cochleariae* zu charakteristischen KKW-Profilen führen.

Phänotypische Plastizität der kutikularen Kohlenwasserstoff-Profile bei Insekten

Neben der experimentellen Arbeit wurde aktuelle Literatur zusammengefasst und diskutiert, die sich mit Studien zur phänotypischen Plastizität von KKW-Profilen bei Insekten in Reaktion auf verschiedene abiotische und biotische Faktoren befasst.

Umfassende Kenntnisse über die Effekte von Umwelteinflüssen auf die KKW-Profile von Insekten sind nötig, um die Mechanismen und Konsequenzen phänotypischer Divergenz zu verstehen. Das Kapitel beschäftigt sich mit drei Hauptpunkten; das Unterkapitel "*Dynamik der phänotypischen Veränderungen*" soll die Frage klären, wie schnell sich der chemische Phänotyp von Insekten durch Umwelteinflüsse verändern kann; zusätzlich werden zwei "*Modi phänotypischer Veränderungen*" analysiert und diskutiert; dabei wird zwischen graduellen und saltatorischen Veränderungen unterschieden. Abschließend werden die "*Auswirkungen phänotypischer Veränderungen*" im evolutionsbiologischen Kontext kritisch diskutiert; hierbei werden v.a. die Auswirkungen phänotypischer Veränderungen der chemischen Kommunikations-systeme herbivorer Insekten auf Artbildungsprozesse berücksichtigt.

Acknowledgements

The work presented in this thesis would absolutely not been possible without the support and help of many persons. Without their kind help I could not have finished my thesis successfully.

First of all, I thank my supervisor Prof. Dr. Monika Hilker for her guidance and mentorship, her patience, encouragement, for providing excellent working facilities and for inspiring and helpful discussions.

I would also like to thank Prof. Dr. Joachim Ruther, who kindly agreed to be the second supervisor on my thesis.

Special thanks go to Dr. Sven Geiselhardt who introduced me not only into the secrets of GC-MS analysis but also into the fascinating world of cuticular hydrocarbons. Moreover, he patiently endured my numerous questions no matter how stupid they were. Without your assistance and many good advices, it would have been much more difficult! I will miss our coffee breaks.

I very much appreciate the help and assistance of all my colleagues and I would like to thank all current and former members of the Department of Applied Zoology/Animal Ecology. In particular, I would like to thank Marc Plevschinski. His work as a bachelor/master student contributed considerably to the realization of this thesis.

Last but always first, a huge hug and kiss belong to my wife for her love, encouragement and endless patience during the PhD period. Lea, I love you!